D1104834

THE RIDDLE OF GRAVITATION

THE RIDDLE OF GRAVITATION

Peter G. Bergmann

Professor of Physics, Syracuse University

Charles Scribner's Sons / New York

In addition to those acknowledgments in the text, the author wishes to thank Ernest and John Bergmann for processing photographs; Margot Bergmann for giving advice on the arrangement and presentation of the material; Helen Dukas and Otto Nathan for helping to select the letter by Einstein and giving permission for its use; Diana Grove for typing the manuscript; and Allan Sandage for selecting and supplying photographs.

Drawings by Ken Fitzgerald

1 3 5 7 9 11 13 15 17 19 B/C 20 18 16 14 12 10 8 6 4 2

Printed in the United States of America
Library of Congress Catalog Card Number 77-82288
ISBN 0-684-15378-5

To the memory of my revered teacher

ALBERT EINSTEIN

Preface

For the past fifty years, Einstein's theory of gravitation, known as the general theory of relativity, has been accepted as the theory which is most intellectually satisfying and which most nearly encompasses all that is observationally known about gravitation. Both scientists and philosophers have discussed the implications of the general theory of relativity for man's ideas about space-time and about the nature of the physical universe. Both scientists and philosophers have responded to the challenge of making the theory accessible to the general public, and even after thirty years, Einstein and Infeld's *The Evolution of Physics* remains a classic of scientific writing addressed to the non-expert.

In this book, I have endeavored to give a fresh presentation of the general theory of relativity. New techniques of astronomical observation have enormously increased man's ability to plumb the depths of the universe. The classical astronomer could observe extraterrestrial objects only through the whole thickness of the atmosphere and in terms of visible light, aided, to be sure, by the telescope and spectroscope. Today's explorer can fly some of his instruments high above the atmosphere; he can make use of radio waves emanating from celestial objects, of cosmic rays and

X rays. Among the fruits of these new techniques is the discovery of new astronomical objects, the quasi-stellar objects (also known as QSOs, or quasars), and the X-ray sources, most of which have not as yet been identified with visible objects. Aside from the intrinsic interest of these exotic objects, there is considerable hope that their investigation may eventually contribute to the understanding of the structure and history of the universe as a whole.

The opening of these new horizons makes me believe that the non-expert needs a fresh introduction to the general theory of relativity. For general relativity represents one of the truly basic approaches to the innermost nature of the physical universe. In this book, I shall try to take the reader through the conceptual foundations of both Newton's and Einstein's theories of gravitation without aiming to give a technical mathematical introduction at the working level. Since understanding the general theory of relativity requires some acquaintance with the ideas of the special theory of relativity (which is not itself concerned with gravitation), the first part of this book is devoted primarily to the special theory. The second part is concerned with the general theory of relativity proper. The third part deals with some recent developments and applications of the theory.

I have tried to present the theory as a creative response to what is presently known about gravitation and also to indicate, in the concluding sections of the book, how the theory may develop in coming decades. Scientists have no better record than political analysts in predicting the future, and I am fully aware of the likelihood that physics may develop very differently from what I foresee. I have attempted to demonstrate that even the "exact sciences" have not reached their ultimate goal in our time, nor are they likely to in the foreseeable future. Not to have grasped the ultimate truth does not frustrate the quest for new understanding. On the contrary, visualizing new frontiers of comprehension depends on a continuing conquest of territory not accessible to our forefathers. All men can rejoice in an undertaking whose success is not predicated on hurting other human beings but whose every step forward is a triumph for all mankind.

PETER G. BERGMANN

Contents

III / RECENT DEVELOPMENTS

APPENDIXES

Illustrations

PLATES

FIGURES

THE RIDDLE OF GRAVITATION

INTRODUCTION

The Scope of Gravitation

A theory of gravitation—the universal attraction that massive bodies exert on each other, the prime mover of the planets about the sun and of the moons about the planets—created by Newton, stands at the cradle of modern exact science. Another theory of gravitation, the work of Einstein, represents one of the proudest achievements of twentieth-century theorizing. Through hundreds, even thousands of years of history, men have observed and measured the effects of gravitation, have endeavored to turn it to their own use, to overcome the limitations imposed by it, and most recently, to control it with extreme precision in the course of their first faltering steps into the depths of space.

Man's continuing quest for knowledge and understanding of surrounding nature has challenged his physical and his intellectual resources, and has caused him to develop both, as has his need for protection from, and control of, this environment. Systematic exploration has led to the emergence of the various branches of the natural sciences, among them physics. It is futile to attempt any formal separation of the jurisdictions of these individual sciences. "Physics is whatever physicists do"; the same holds of chemistry, biology, astronomy, and the other natural sciences. In trying to understand the dynamics of matter, physicists have come to recognize that it is composed of atoms and molecules and that the behavior of materials, such as crystals or gases, can be best understood in terms of the interplay of the constituent atoms. Atoms in turn are complex systems, composed of nuclei and of negatively charged particles, the *electrons.* Finally, atomic nuclei consist of two kinds of building blocks, *protons,* which are positively charged, and *neutrons,* which have no charge.

The enormous complexity of actual matter is largely the result of this multilayered structure, whose ultimate constituents, the elementary particles, exhibit but a few modes of interaction. These modes of interaction involve vastly different powers. The particles making up atomic nuclei are bound together by the most powerful of these forces, and enormous expenditures of energy are necessary to separate them from each other. Such energies are produced typically in large machines whose power is expressed in terms of millions and billions of volts. The electrons within atoms are bound to the nuclei by electromagnetic forces; application of quite moderate energies, a few volts—energies like those encountered in chemical reactions—is sufficient to separate electrons from the nuclei. The weakest interactions of all are those of gravitation, as long as one thinks in terms of elementary particles and of atoms.

To get an idea of the relative strengths of electric and gravitational forces, consider the interactions between two electrons, the particles that are the carriers of the electric current in storage batteries and in electrical wiring. To reduce their electric interactions to the level at which the same electrons attract each other gravitationally at a distance of one-hundredth of an inch, one would have to separate the two electrons by a distance of fifty light years, roughly ten times the distance between neighboring stars, or some thousand trillion (1,000,000,000,000,000 or 10^{15}) miles (Appendix VI).

In terms of the interactions between elementary particles, gravitational forces are weak almost beyond imagination. Nevertheless, they alone determine the motions of celestial bodies. For gravitation combines two characteristics that tend to reenforce its effects when large bodies are involved. In contrast to the binding forces within atomic nuclei, whose reach does not extend even to distances of the order of the diameter of an atom, gravitational binding forces remain significant at large distances. Further, whereas within and between atoms there are both attractive and repulsive electric forces, which tend to neutralize each other for large, electrically uncharged bodies, all gravitational forces are attractive; that is, bodies invariably gravitate toward each other. Hence, gravitational forces, and their effects, are the only ones that play any role in determining the paths in which the members of our solar system revolve around the sun.

Celestial bodies move about in the vast emptiness of space unhampered by the many complications that beset the motions of objects in everyday experience. They encounter neither the friction of the ground nor the resistance of the atmosphere. The celestial bodies act on each other while separated by large distances. These are among the reasons explaining why the dynamics of celestial bodies was disentangled so early in the history of modern science. With astronomers like Galileo and Kepler having laid much of the groundwork, Newton succeeded in fashioning a single theoretical structure and in casting it into a mathematical form. His theory satisfied man's desire to understand rationally the trajectories on which the celestial bodies travel in space, and enabled astronomers to predict their movements with great precision.

With improved techniques of observation and with ever more

powerful methods of computation, the successes of Newton's theory of gravitation mounted steadily over a period of some three hundred years. His work has remained a standard against which to judge the validity and success of a physical theory. About fifty years ago, Newton's work was modified, and partially superseded, by a new theory of gravitation, Einstein's general theory of relativity. This new development was needed not because of conspicuous failures of Newton's theory of gravitation but because of inconsistencies between it and requirements of electromagnetic theory, which Einstein had revised earlier in his special theory of relativity.

Although the general theory of relativity has explained a few minor descrepancies between astronomical observations and Newtonian theory, its most important contribution to date has been a complete revolution of our notions of space and time, a revolution that may not yet be completed. Immediately after the publication of Einstein's theory of gravitation in 1916, there was a great deal of excitement, surrounding particularly the observations that tended to bear out the validity of Einstein's proposals. Then, for several decades, active work in general relativity was confined to a rather small group of physicists and mathematicians, who remained aware of the potentialities of the new theory and who proceeded to explore and to develop it further. Since the 1950's the number of active workers in general relativity has rapidly increased. Renewed interest has been brought about by recognition of the conceptual power of the theory and of its possible bearing on other areas of physics, by new possibilities of experimental and observational work in this area, and no doubt also by its possible implications for space travel and space exploration.

I / NEWTONIAN PHYSICS AND SPECIAL RELATIVITY

1 Early History

As far back as history records, men have known that a small number of celestial objects moved across the heavens differently from all the other stars, which travel together on periodic orbits. Aside from the sun and the moon, the planets Mercury, Venus, Mars, Jupiter, and Saturn were known to travel along individual paths. These motions aroused the interest of navigators and calendar makers, the planets were associated with religious and mystical ideas, and in several civilizations men tried to evolve schemes for predicting their movements. According to the Hellenistic astronomer Claudius Ptolemy (127–141) the earth formed the center of the universe. As long as the Ptolemaic view prevailed, planetary orbits were considered identical with their apparent movements across the sky; explaining them required spheres moved by gears within gears in order to account for the observed bends and loops.

In the seventeenth century the German Johannes Kepler (1571–1630) organized the painstaking observations of his Danish predecessor, Tycho Brahe (1546–1601), on the basis of the view of Nicholas Copernicus (1473–1543). The Polish astronomer declared that the sun rather than the earth forms the center of

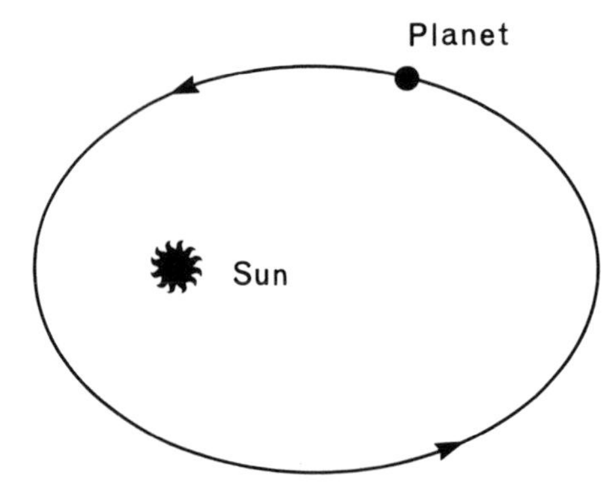

Fig. 1. Orbit of a Planet, with Sun at Focus

planetary motions. Kepler succeeded in showing that the planets move along elliptic paths and that the sun lies at a *focus* of each of these *ellipses* (Fig. 1). Each planet moves so that a straight line drawn to connect it with the sun sweeps out equal areas in equal times. The planet in Fig. 2 takes the same time to move from C to D as from A to B. Finally, the periods of revolution of the different planets are proportional to the 3/2 power of the largest diameter of each orbit. Figure 3 shows Kepler's 3/2 power law in graphic form, with both diameters and periods of revolution plotted logarithmically in order to accommodate the req-

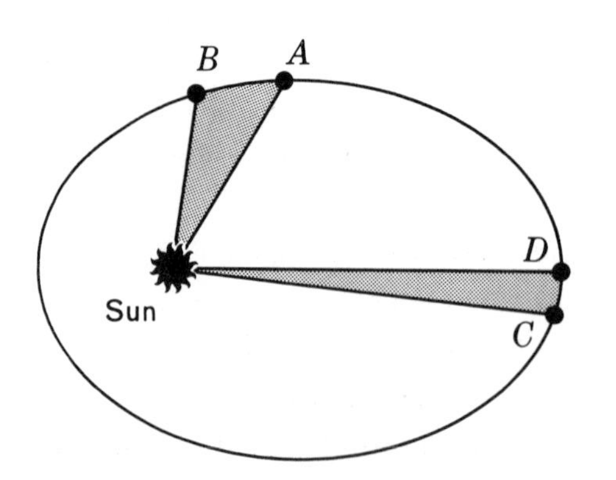

Fig. 2. Equal Areas Law

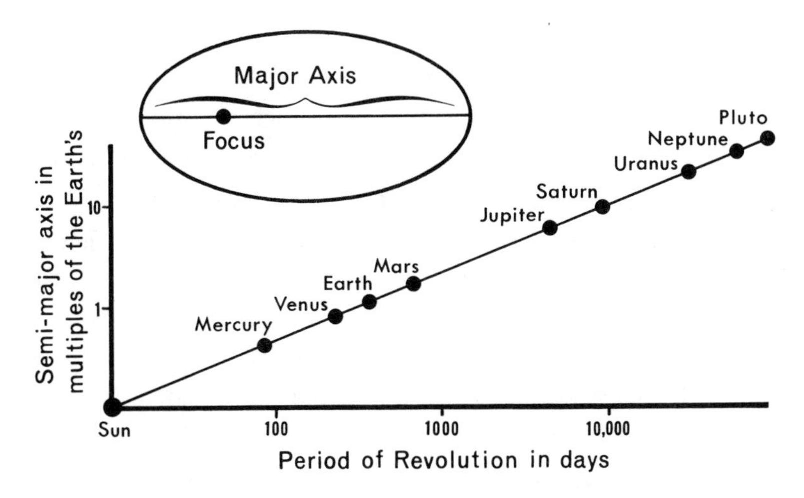

Fig. 3. Kepler's Law concerning Periods of Revolution

uisite ranges—Mercury's period of revolution is 88 days; Pluto's, 91,000 days. The straight line represents the 3/2 power law. The data pertaining to the individual planets are represented by small circles.

The discoveries just mentioned enabled Newton to formulate the laws of mechanics in general and those of gravitation in particular. It is almost impossible for us to do full justice to the genius of the British mathematician and physicist Sir Isaac Newton (1642–1727). He was able to develop Kepler's laws into a comprehensive physical theory only because he managed first to create the necessary mathematical tools: Newton "invented" *differential* and *integral calculus*, the basic mathematical techniques required for dealing with variable quantities, such as the movements of bodies in the course of time. By means of the techniques of calculus, he succeeded in drawing from Kepler's empirical laws the principles of motion that applied every instant of time and thus shaped planetary motions into complete orbits.

If the time rate of motion, or *velocity,* of a planet is known at all times, then by employing differential calculus, one can compute at every instant the rate of change of velocity, or *acceleration.* Conversely, through integral calculus, one may compute, from a knowledge of velocity, the distance traversed in a finite given time. Newton recognized that Kepler's law specifying that the line connecting planet and sun sweeps out equal areas during equal times implied that any change in the velocity of a planet could be directed only toward the sun, not at right angles to it (Appendix I). Newton reached three conclusions, one already divined by the Italian Galileo Galilei (1564–1642), the other two his own. According to the first, the action of the sun causes the planets to change their velocities continuously, and this change, the acceleration (rather than velocity itself, or location, or total trajectory) ought to be the subject of relatively simple laws. In this respect, Newton eliminated the medieval notion that each body had a proper *location* in the universe, and that the action of other bodies caused a *displacement.* In his view, a body not acted on by others would continue to move at the velocity it happened to have acquired at some earlier time; the result of action by other bodies was to cause a change in that velocity, an acceleration.

Second, as mentioned earlier, from the equal-areas law of Kepler, Newton concluded that the force exerted by the sun on each planet was directed toward the sun, that it was an *attractive* force, as illustrated in Fig. 4. As the planet proceeds in this drawing from location P_1 to location P_2, its velocity changes from v_1 to v_2. The change, Δv, is directed toward the source of the force, the sun. The velocities, v_1 and v_2, are to be understood as *vectors* (directed quantities). The change, Δv, is the vector leading from the head of arrow v_1 to the head of arrow v_2, when their tails are placed at the same point.

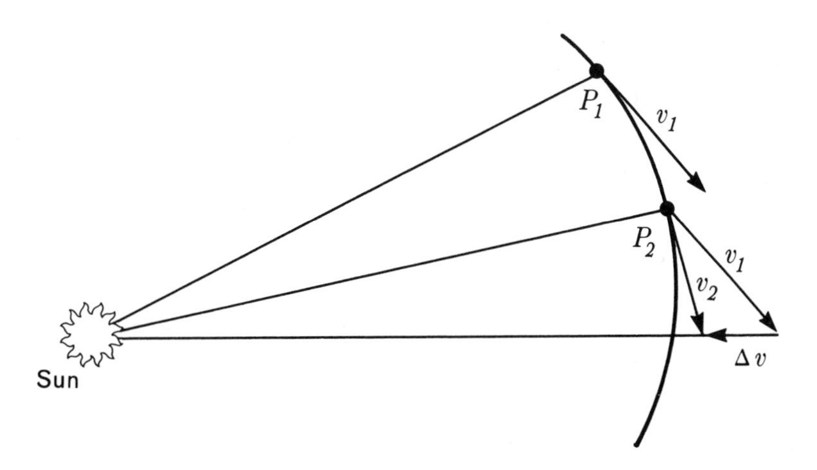

Fig. 4. Gravitation Is an Attractive Force

Finally, both the shapes of the planetary trajectories and the 3/2 power law governing the periods of the planetary orbits led Newton to the conclusion that the strength of gravitational attraction between any two bodies depends on their distance from each other, and that this strength is inversely proportional to the square of the distance (Fig. 5): If one could measure the strength of gravitational attraction between two bodies one million miles apart, and again when they were two million miles apart, one should find that, at the greater distance, the attractive force has dropped to one-fourth of its former value (Appendix II).

In Newton's time, only two kinds of force were available for quantitative investigation. One was the force of gravity; the other the forces of push and pull encountered in everyday life, pushing a baby carriage, say, or pulling a dog by its leash. Newton endeavored to construct a general theory of all forces, both those known in his time and those that might be discovered and investigated later. He intended his theory of gravitation to be the one example that he himself could work out fully in all its rami-

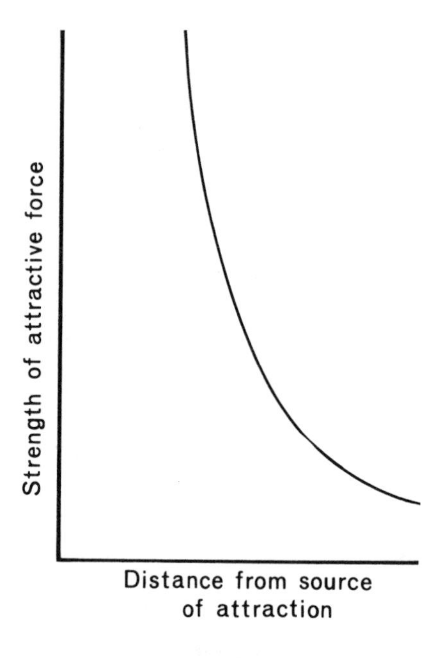

Fig. 5. The Inverse-Square Law of Gravitational Attraction

fications. As for the general theory of forces, Newton formulated his celebrated three laws: (1) In the absence of force, a body will continue at rest or in its present state of uniform rectilinear motion. (2) In the presence of force, a body will be accelerated in the direction of that force, the product of its mass by its acceleration being equal to the force ($f = ma$). (3) To every force there corresponds an equal counterforce, acting in a direction opposite to that of the force (Fig. 6). According to the third law, then, each planet exerts an attractive counterforce on the sun, accelerating it toward the planet, but producing a relatively small acceleration, because the mass of the sun so vastly exceeds the mass of every planet in the solar system.

Among all possible existing forces, the force of gravity may be characterized first of all by its universal occurrence. All bodies possess *mass,* defined as the ratio between force applied to them and the resulting acceleration. The attractive gravitational force between any two bodies depends on both their masses, being proportional to the product of the two masses involved. Moreover, the force of gravity is governed by the inverse-square law (Fig. 5). Other forces might conceivably depend on distance in a different manner; today we know that some do.

One aspect of the universality of gravity, the peculiar double role played by the mass, was to become a cornerstone of Einstein's general theory of relativity. According to Newton's second law, mass is an attribute of every body that determines how it is to yield to a force brought to bear on it, whether this force be gravitational or of some other kind. As all bodies, according to Newton, respond to an external force by undergoing acceleration (change of velocity), the mass of a body determines how much acceleration will result from the application of a fixed force. If the same effort is expended to push a bicycle and an automobile, the two will not accelerate from rest to the same velocity in the same time. The more massive object, the automobile, will be set in motion much more slowly. Conversely, to force the bicycle and the automobile to undergo the same change of velocity one must push the automobile much harder (exert a much larger force on it) than the bicycle.

But in the field of gravity, mass plays a second role, quite

Fig. 6. Newton's Third Law

distinct from its being the ratio between force and acceleration: Mass also serves as the source of gravitational attraction; at a given distance, the pull that two different large bodies exert on the same small object depends on their respective masses. In fact this pull is proportional to the mass of the source. Likewise, because of Newton's third law, the attractive forces that two different bodies experience under the influence of the same source of attraction (and at the same distance from it) is proportional to their respective masses. In engineering science, and in everyday usage, the force with which any body is attracted to the earth is called that body's *weight.* If an individual were to find himself in outer space, far from earth and all other sources of gravity, he would find himself weightless, though by no means massless. No doubt, space medicine will in time develop diets for space travelers that will help them to control their bodies' masses, though weightless they will be in any case.

To summarize: Mass enters into the relationship between force and acceleration; mass also determines the magnitude of gravitational attractive forces. These twin roles of mass cause the accelerations of different bodies in the same gravitational field to be the same: Given two different bodies of masses m and M, respectively, both permitted to fall freely toward the earth, the ratio of the attractive forces experienced by these two bodies will equal the ratio of their respective masses, m/M; but their resulting accelerations will be equal. Consequently, the gravitational acceleration common to all bodies in the same gravitational field is independent of the characteristics of the falling bodies. It depends only on the masses and distributions of the bodies causing that force. The dual role of mass, and the resulting equality of gravitational accelerations of all bodies in the same gravitational field, is called the *principle of equivalence.* This name has a historical significance in that gravitational and inertial effects (to be discussed later) are to some extent equivalent. But it is safer today to let the term principle of equivalence denote only the equality of mass as a measure of a body's resistance to acceleration and of mass as a source of gravitational attraction.

On the surface of the earth, the universal gravitational acceleration amounts, in very rough figures, to 32 ft/sec^2, or 10 m/sec^2 in metric units. The speed of a free-falling object, if the resistance of air to the fall is unimportant, increases by 32 feet per second every second. For instance, starting from rest, its speed will be 96 ft per sec by the end of the third second of unrestrained free fall. (In metric units, 32 ft equal approximately 10 m.) This rate of increase in speed is commonly designated in elementary physics textbooks by the symbol g. Because the earth's shape is not precisely spherical, g does not have exactly the same value everywhere. It is larger near the poles than near the equator, and smaller on top of a high mountain than in a deep valley. If g is determined with sufficient precision, it is found that even geological structure has an effect on its value; for this reason, the geophysical methods used in prospecting for oil and other mineral resources include precision determinations of g.

That at a given location all bodies undergo the same acceleration is a peculiarity of gravity, which is not shared by other kinds of force. For instance, in an electric field, particles with different charge-to-mass ratios are accelerated differently. A physical instrument, the mass spectrograph, is used to sort out particles by this means; commercial models usually employ a combination of electric and magnetic fields to effect the separation of different kinds of particles. It is out of the question to design a mass spectrograph with purely gravitational forces, because in it all particles, regardless of their properties, would accelerate exactly alike.

Although Newton could do no more than record the fact, he did know about the universality and uniformity of gravitational acceleration. It was left for the German theoretical physicist Albert Einstein (1879–1955) to interpret the underlying principle, the principle of equivalence, so as to make it the cornerstone of our current ideas about the nature of space and time.

About the turn of the century Roland von Eötvös, a Hungarian physicist (1848–1919), tested the principle of equivalence experimentally with an accuracy of one part in one hundred million and confirmed it. Recently, Robert H. Dicke of Princeton University has repeated Eötvös' work and improved on his accuracy by another factor of a thousand. Again, Dicke's tests have confirmed Newton's original assertions, now to an accuracy of one part in one hundred billion (10^{-11}).

2 Relativity of Motion

Up to this point I have treated velocity and acceleration of particles as if these aspects of bodily motion called for no qualification, as if everybody understood motion and its measurement without further ado. But motion, or for that matter the absence of motion, *rest,* is unambiguous only as long as there is general agreement that motion is to be referred to the earth. As long as one is concerned primarily with events taking place on earth and in everyday surroundings, this attitude is eminently justified. If the law prescribes a speed limit, say, fifty miles per hour, on a particular highway, drivers, police, and traffic court are all agreed that the speed is to be reckoned with respect to the pavement, which in turn is firmly connected to the earth as a whole.

When it comes to airplanes, one ordinarily specifies speed as either "air speed" or "ground speed," and the two are not equal (Fig. 7). *Air speed* is the speed of the plane relative to the surrounding air mass; its magnitude depends on the performance of the motors, on the shape of plane body and wings, and the like. *Ground speed* is the speed relative to the earth's surface; it determines the time that it will take the plane to travel the distance between two specified airports. The difference between

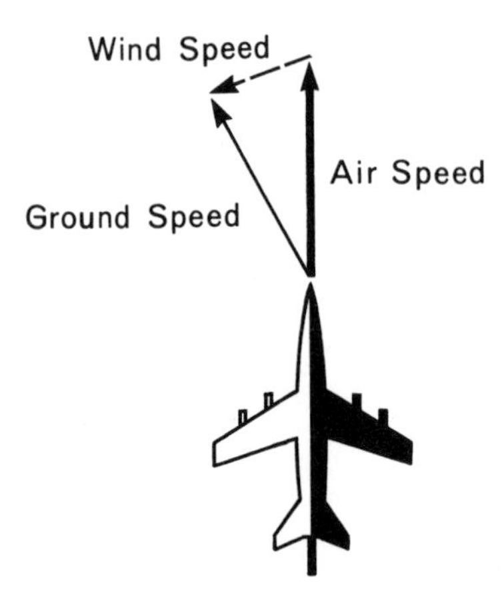

Fig. 7. Air Speed and Ground Speed of a Plane

air speed and ground speed depends on the amount of head wind or tail wind encountered. The terms air speed and ground speed, then, indicate to which reference object the plane's motion is to be related, the atmosphere or the solid ground.

During medieval times, men accepted Ptolemy's view that the earth was the natural center of the universe. In this context, adopting earth as a universal *frame of reference* (standard to which all motions are referred) was justified not only in everyday living, but also in fundamental science. Once the Ptolemaic point of view was abandoned, the choice of a frame of reference was reopened. Copernicus substituted the sun for Ptolemy's earth as the "natural" frame of reference, and his choice was indeed excellent for describing the motions within the solar system. Today, however, it is understood that the sun is but one of millions of fixed stars in the galaxy, and further, that our galaxy is but one of innumerable galaxies visible through a large telescope.

Newton was well aware of the profundities involved in the choice of a proper frame of reference. All his fundamental laws of mechanics involved statements concerning accelerations, changes in the velocities of physical bodies, rather than the velocities themselves. These accelerations were tied to distances between the bodies, such as the distance of the earth from the sun, of the moon from the earth, and the like. The choice of a frame of reference had no effect on the determination of distances of objects from each other, but the accelerations which resulted from the mutual attractions and repulsions of bodies were to be reckoned in relation to a universal norm, a norm intimately bound up with the choice of a frame of reference.

Anyone who rides in an automobile or flies in an airplane knows that as long as ride or flight is straightforward and at constant speed, one adopts, almost instinctively, the vehicle as one's frame of reference. Indeed, while a plane takes the passenger at several hundred miles per hour toward his destination, he may walk down the aisle at one mile per hour toward another passenger without being much aware that he continues to travel at a ground speed that is barely changed by his own movements. But during takeoff, for instance, the passenger is pressed hard against the back of his seat; and no one thinks that this is so because the passengers are temporarily "attracted" by the plane's tail. Rather, during takeoff or landing, the body of the plane is a conspicuously "unsuitable" frame of reference, with respect to which bodies will exhibit accelerations not accounted for by forces caused by nearby objects. In short, only a "suitable" frame of reference will not distort "true" accelerations.

It remains to determine more precisely which frames of reference are "suitable." Consider two observing scientists, one firmly connected with the earth, the other with a vehicle moving at constant speed along a straight-line track. If both observers collected data on the flight of the same bird, they would find its speed at any given instant to be a matter of dispute, as each observer would refer the bird's motion to himself as the standard;

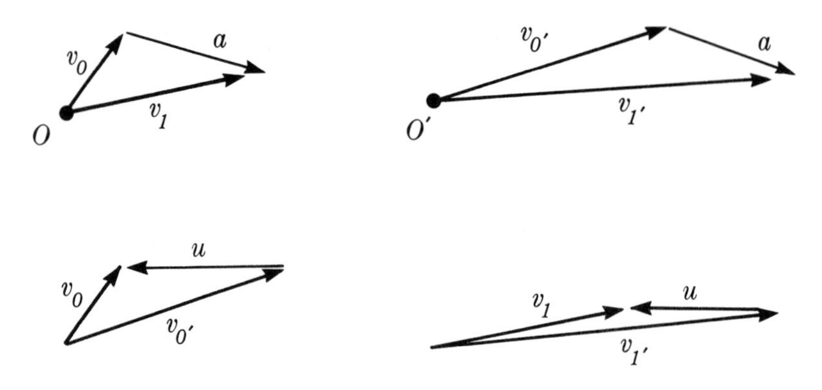

Fig. 8. Acceleration of a Bird

but they could agree concerning the bird's change in velocity (change in speed as well as direction of flight). Initially (Fig. 8), the observer on the ground determines the bird's velocity to be v_0; 1 second later, it is v_1. He finds its acceleration, also a vector, to be a. The moving observer's own motion relative to the ground is u; hence he observes the bird's motion as v_0' and v_1', respectively; but the change from v_0' to v_1' is a, the same as the change from v_0 to v_1.

If the laws of nature are concerned only with changes in velocity, as Newton assumed, both observers are equally capable of obtaining the data that would permit testing Newton's laws of dynamics. Consider now a third observer O'' who is to be attached to an automobile careening along a curved race track. Changes in velocity as recorded by him will deviate grossly from those determined by the two previous observers. The ground velocity of the third observer changes from w_0 (Fig. 9) to w_1 in the course of 1 second. Accordingly, he determines the bird's acceleration to be a'', which differs from a. Thus one is led to conclude that, for collecting data relevant to an experimental confirmation of Newton's laws of mechanics, one may consider equivalent all observers who, relative to one another, are engaged in straight-line and unaccelerated motion.

There remains the question whether, among all conceivable observers, one class is to be considered preferentially qualified to collect data on the dynamics of various bodies and physical systems. Newton answered this question affirmatively. According to his first law, a body free of external forces (of the influence of another body) should experience no acceleration; it must be possible to find an observer for whom this statement is valid. Such an observer will be called an *inertial* observer; relative to him, the motion of a forcefree body will be unaccelerated. If an inertial observer is considered the hub of a scaffolding, consisting of surveyor's rods, or of equivalent optical instruments (such as transits), one calls the whole framework an *inertial frame of reference,* or for short, an *inertial frame.* The existence, and equal standing, of a set of inertial frames—all of which must be at rest or in uniform straight-line motion with respect to each other—is thus a fundamental tenet of Newton's theory of the physical universe. The equal validity of all inertial frames of reference, and the non-existence of one frame representing absolute rest, is

Fig. 9. An Accelerated Observer

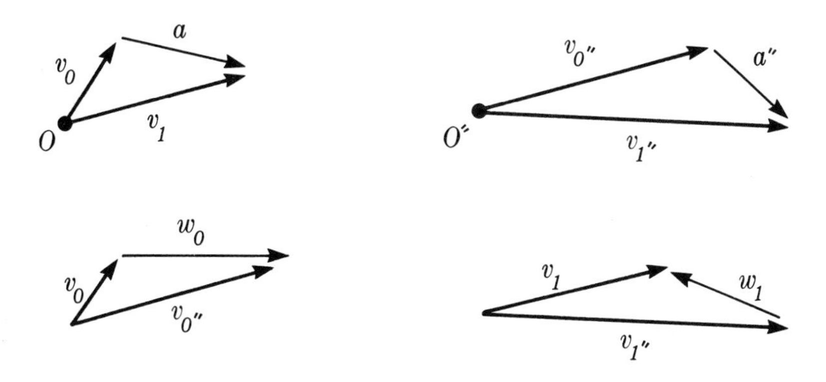

known as the *principle of relativity*. The principle of relativity remained unquestioned for about two hundred years. During this time physicists accommodated themselves to the notion that there was no such thing as absolute rest, or absolute motion for that matter, but only absolute acceleration, and that the absolute acceleration of physical bodies was governed by the forces resulting from the proximity of other bodies.

3 The Universal Speed of Light

In order to define *absolute rest* and *absolute motion,* one must be able to single out from among all the inertial frames one frame of reference not merely unaccelerated but having some additional special quality (not possessed by the other inertial frames) that would render it *the* frame of choice for describing all processes in the universe. And presumably, there was nothing in the universe conceived by Newton that would have furnished a means for selecting this one frame.

During the nineteenth century, the science of electricity and magnetism was developed by a number of physicists: the Dane, Hans C. Oersted (1777–1851); two notable Englishmen, Michael Faraday (1791–1867) and James Clerk Maxwell (1831–1879); and Heinrich Hertz (1857–1894), a German. Maxwell in particular noted that the phenomena of electromagnetism did not fit into the scheme of Newtonian mechanics. Whereas it had been thought that only the distance between two objects determined the force one exerted on the other, electric charges in motion, such as are met with in electric currents, were found to produce effects not encountered when charges are at rest. Celestial bodies will only attract each other; electric charges at rest will either attract or repel each other, but in any case they will exert forces only in the direction of the connecting straight line. Oersted discovered that an electric current (which consists of electric charges in motion) will exert a force on a magnetic needle at right angles to the connecting straight line. Previous observations in astronomy had tended to show that the force between two bodies depended only on their instantaneous configuration, but Hertz showed by experiments that electromagnetic disturbances propagate as waves, at a finite rate of speed. Hence the force experienced by one body can be understood and explained only in terms of the history of the other.

Maxwell succeeded in casting all known electromagnetic effects into a mathematical form that has endured to this day. The whole theory is usually compressed into a small number of equations, known as *Maxwell's field equations.* Based on Faraday's earlier work, Maxwell's stressed the notion of *fields,* in contrast to Newton's emphasis on the direct action of bodies on each other across empty space (*action at a distance*). Faraday and Maxwell regarded the effect of an electrically charged body as giving rise to stresses in its immediate surroundings. These in turn produce stresses in ever widening circles, gradually diminishing as the distance from the original source increases. These stresses, which are to be thought of as capable of existence in otherwise empty space, are called fields. Particles, according to Faraday and Maxwell, experience forces in the presence of fields; thus the fields are intermediaries between the material particles and assume the burden of Newton's action at a distance.

Maxwell's field equations describe mathematically the relationships between electric charges and currents and the resulting electric and magnetic fields and the relationships between the electric and the magnetic fields themselves. For instance, one set of Maxwell's equations is to the effect that, in the presence of a magnetic field which changes in the course of time, an electric field arises which is not caused by the presence of any electric charge. This *law of electromagnetic induction* is utilized in electric generators, which induce electric voltages in the winding of their armatures by subjecting them to varying magnetic fields.

From his theory, Maxwell not only predicted that electromagnetic fields propagate at a finite rate, but he also determined the speed of propagation, 186,000 miles per second, the same as the speed of light. Hence he conjectured that light was a species of electromagnetic phenomenon, a conjecture subsequently confirmed. Hertz, influenced by Maxwell's predictions, performed the laboratory experiments that established the existence of electromagnetic waves and that, incidentally, form the point of departure for today's wireless communication technology.

The laws of mechanics involve only accelerations, not velocities; the laws of electromagnetism involve a universal velocity, the speed of propagation of all electromagnetic waves in empty space. Its value, predicted by Maxwell's theory and since confirmed universally by experiments and observations, is the same for all wavelengths, radio waves, light, and X rays.

If there is such a thing as a universal speed, the results of Newtonian physics concerning the choice of frames of reference must be reviewed. As long as the laws of physics were concerned only with accelerations, all inertial frames of reference were equally appropriate for a description of nature. In fact, within the framework of mechanics, no conceivable experiment or observation would lead to the selection of one particular frame of reference as fundamental. But if in empty space light propagates at the universal speed of 186,000 miles per second (from now on this universal speed will be denoted by the symbol c), then a careful determination of the apparent speed of light relative to

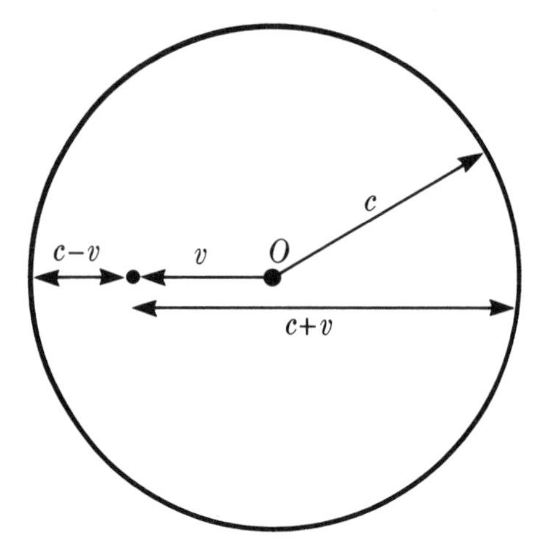

Fig. 10. Apparent Speed of Light according to the Ether Hypothesis

laboratory apparatus should reveal the velocity of that apparatus relative to the rest of the universe. There should exist one frame of reference with respect to which light does travel everywhere at the speed c. Call this particular frame of reference the *frame of absolute rest.* (Historically, this frame has been related to a hypothetical medium of propagation of electromagnetic radiations, the *luminiferous ether.* The ether was thought to play a role in the propagation of light comparable to that of the atmosphere in the propagation of sound.) Then, with respect to any other frame of reference, the apparent speed of light should be less than c in the direction in which this frame is traveling relative to the frame of absolute rest; it should be greater than c in the opposite direction. In Fig. 10, the circle represents the spreading of a light signal from point O during the first second. If, during the same second, an observer moves the distance v to the left, then relative to him, the light signal should have traveled to the left only a distance $(c - v)$ but to the right as much as $(c + v)$.

To discover the state of absolute rest would appear to call for precision determinations of electromagnetic phenomena. Forthwith, a number of experimentalists proceeded to design experiments accomplishing this end, among them the American physicists Albert A. Michelson (1852–1931) and Edward W. Morley (1838–1923).

I shall not dwell on the details of these investigations. It is enough to say that much ingenuity went into attempts to discover deviations of the speed of light from uniformity in all directions; these efforts were designed with an accuracy sufficient to discover even a minute effect of the kind envisaged. All these attempts failed; they led to the result, hard to accept at the time, that there was no absolute motion of the earth with respect to the privileged frame of reference. The Dutch physicist, Hendrik Antoon Lorentz (1853–1928) contrived a theoretical scheme according to which absolute motion of physical objects, including measuring rods, should compress them in such a manner that differences in the speed of light remained undetectable by any conceivable apparatus. Jules Henri Poincaré (1854–1912), the French mathematician, suggested that the consistent failure to identify the frame representing absolute rest indicated that no such frame existed, and that Newton's scheme of the multiplicity of inertial frames was valid after all. In 1905, Einstein combined Lorentz' and Poincaré's ideas into a new approach to the issue of frames of reference and so was able to explain why no experiment had uncovered the absolute motion of the earth, without contradicting Maxwell's theory of electricity and magnetism. Einstein realized that, by modifying traditional views on the nature of space and time, one could, without contradictions, accept both the equivalence of all inertial frames of reference and the validity of Maxwell's theory.

4 The Special Theory of Relativity

Ever since Newton's time, it had been accepted that all frames of reference are representable by individual frameworks of rods or other systems of markers identifying locations in space but that all observers "naturally" make use of the same universal time. This assumption is so intuitive that usually it was not even stated explicitly but simply taken for granted. As long as one is concerned with everyday happenings on earth, the assumption is borne out by experience. No one would question that the timing of a race by the umpires would produce the same results as timing by the participating contestants, provided that everyone was honest and that all clocks used were mechanically in good condition.

Einstein, however, pointed out that comparing clocks regardless of their motion was simple only as long as all velocities involved were very much smaller than the speed of propagation of light—about 186,000 miles per second (300,000 km/sec in metric units). As no method of signaling is known that transmits information faster than light, new problems arise when one needs to compare clocks whose relative speed is, say, 100,000 miles per second. For, to compare the rates of two clocks, one must take at least two readings of both; otherwise one cannot be certain that their respective rates are equal. Suppose now that one sets two clocks so that they indicate precisely the same time as they pass each other, a second reading would have to be taken when they have moved a considerable distance apart. How can one then be certain that on this second reading the two clocks are observed *at the same time?* One could provide this assurance by specifying that the reading is to be taken by a third observer who will be exactly midway between the two clocks at the instant the second reading is taken. But a detailed analysis, not reproduced here, shows that such an instruction is ambiguous: However phrased, such instructions will lead to different results if the roles of the two clocks are interchanged.[1]

The first result of Einstein's analysis was the discovery of the *relativity of simultaneity:* two events taking place at widely separated locations may appear to occur at the same time to one observer, but at different times to an observer in a different state of motion. Accordingly, the assumption of a common, or universal, time cannot be justified on the ground that using a fixed procedure would enable any observer to establish such a universal time independently of his state of motion. Rather, one must assume that each observer could construct his own frame of reference; such a frame of reference should consist not only of a framework of surveyor's rods to identify locations in space but also of clocks placed at diverse locations, all traveling together with the observer and synchronized to his satisfaction. If a second observer in a different state of motion viewed the clocks be-

[1] Albert Einstein, *Relativity, the Special and the General Theory: A Popular Exposition*, trans., R.W. Lawson (London: Methuen & Co., Ltd., 1954).

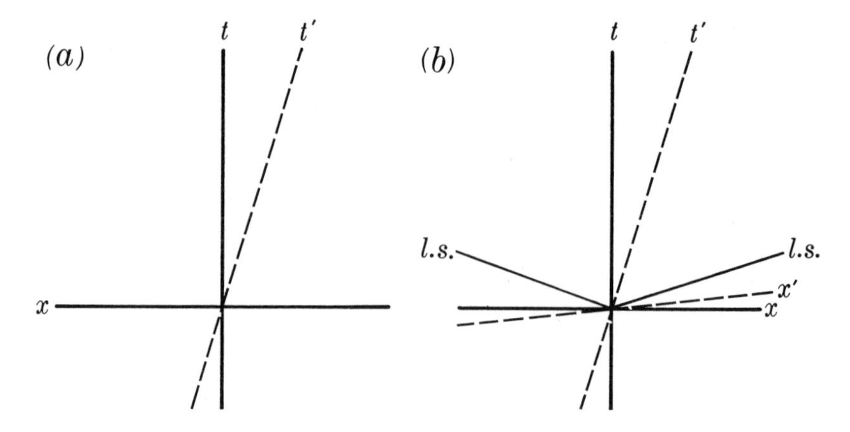

Fig. 11. The Lorentz Transformation

longing to the first frame, he should find that they are not synchronized (though all going at the same rate): To him the clocks located forward in the direction of the motion of the first frame would appear relatively retarded, and those located backward would seem advanced.

As his next step, Einstein set up a new system of relationships between time and distance measurements in two distinct inertial frames of reference. In accordance with experiment, he required that the speed of propagation of light be the same in terms of the data obtained by either observer, even though they performed their respective determinations on the same light pulses. If time be considered universal, then the horizontal line, x, connects all locations (x) at the same time ($t = 0$), for all observers (Fig. 11a). The vertical line, t, connects all instances in time (t) at the same location ($x = 0$) for one observer; the slanting line, t', for the other observer. If the second observer is to see the speeds of propagation of light signals (l. s.) as equal in both directions (see t'-axis, Fig. 11b), his line of equal t'-time (marked x') must be at a slant relative to the line x.

The system of relationships just described is called a *Lorentz transformation*. Einstein showed, further, that if these newly postulated relationships are valid, then Maxwell's laws of the electromagnetic field can hold fully in both inertial frames.

Thus he resolved the paradox between the system of equivalent inertial frames of Newtonian mechanics and the one frame of absolute rest apparently required by Maxwell's theory. The special theory of relativity once more treats all inertial frames of reference as equals; to this extent, the theory reestablishes the principle of relativity dating back to Newton and the seventeenth century.

Just as universal time had to be sacrificed to the Lorentz transformation, the notions of a universal *rate* of clocks and of a universal length of yardsticks dropped by the wayside. Whereas the term *principle of relativity* originally applied to the relative equivalence of inertial frames of reference, soon physicists and philosophers both began to talk of the "relativity of time" and the "relativity of length." Given two complete frames of reference moving relative to each other, each observer finds the rate of the clocks of the other observer slower than that of his own clocks, and he finds the yardsticks used by his colleagues contracted in the direction of relative motion, though unaffected at right angles to that motion. (See Appendix III for mathematical details.)

Actually, the special theory of relativity replaces "absolute length" and "absolute time," which had to be abandoned, by a new "absolute," often called the *invariant* (or *proper*) *space-time interval*. Given two events, occurring at distant locations, their spatial distance is not absolute, even in the Newtonian scheme of things, only the lapse of time between them. For, unless the two events are simultaneous, an observer moving, with his whole frame of reference, in one way may travel a distance between the occurrence of the two events, so that to him they appear to have taken place at the same location, even though they appear far apart to an observer who has traveled in another direction.

As for the special theory of relativity, the universality of the speed of light (in empty space) imposes on the relationship between any two events A and B a characteristic that must be independent of the inertial frame of reference in which the time interval between them, t, and the distance between them, L, are measured. Suppose that event A involves the emission of a flash of light which just manages to reach the site of B as that event occurs. In that case, t and L are related to each other by the condition

$$L = ct.$$

Such a relationship between the two events is usually called *lightlike.* If, in another frame of reference, the time lapse has been found to be t', and the distance between the two events L', then, because both observers would find the speed of light to be the same, c, L', and t' again must be related by an equation

$$L' = ct'.$$

If two events are in a lightlike relationship to each other for one observer, they will be lightlike to each other for any other observer.

Of course, the time lapse between two events may be too great for the lightlike relationship to hold, that is,

$$L < ct.$$

In that case, the signal emanating from the earlier event will pass the site of the later event before that event takes place. Again, this relationship cannot depend on which observer is charged with performing the appropriate measurements, and any other observer will find that for him, too,

$$L' < ct'.$$

Such a relationship is called *timelike,* because it is possible to find an observer moving in such a fashion that for him the two events take place at the same site, one after the other.

Conversely, if the distance is so large and the lapse of time so short that a flash of light issuing from either event will reach the site of the other event only after that event has come to pass,

$$L > ct$$

then the same relationship must hold for any other observer,

$$L' > ct'.$$

Such a relationship is called *spacelike.* Incidentally, the symbols $<$ and $>$ signify the relationships "less than" and "greater than," respectively.

The invariant space-time interval is related to the lightlike, timelike, or spacelike relationship between two events. In order to understand its significance, it is useful to look at the expressions for the Lorentz transformation (Appendix III). If the time lapse between two events be designated by t, as before, their spatial relationship must be separated into the distance between them along the line of relative motion between the two observers under consideration (a distance marked by the symbol x), and the distance between the two events at right angles to the direction of motion (designated r). The total distance between the two events, L, is related to the two quantities x and r through the Pythagorean law,

$$L^2 = x^2 + r^2$$

and an identical relationship holds between the quantities L', x', and r' measured by the other observer. In terms of these quantities, two events will be lightlike relative to one another if

$$c^2t^2 - x^2 - r^2 = 0.$$

The relationship will be timelike if the left-hand side is positive, spacelike if it is negative.

This sum and difference of squares turns out to take the same value in terms of either primed or unprimed coordinates. The two sets of coordinates are tied to each other by the equations of the Lorentz transformation (Appendix III),

$$x' = \left[1 - \left(\frac{v}{c}\right)^2\right]^{-1/2} (x - vt)$$

$$t' = \left[1 - \left(\frac{v}{c}\right)^2\right]^{-1/2} \left(t - \frac{vx}{c^2}\right)$$

$$r' = r.$$

With the help of these relations, by straightforward substitution, one verifies that, for the two observers, one always has

$$c^2t^2 - x^2 - r^2 = c^2t'^2 - x'^2 - r'^2 \quad \text{or}$$
$$c^2t^2 - L^2 = c^2t'^2 - L'^2.$$

Accordingly, one can introduce, for timelike relationships between two events, the *proper time interval* between them, T, by the definition,

$$T^2 = t^2 - \frac{L^2}{c^2} = t'^2 - \frac{L'^2}{c^2}.$$

This interval T will go to zero if the relationship between the two events changes from timelike to lightlike. If the relationship is spacelike, however, T is no longer a real quantity (T^2 becomes negative), and it is replaced by the *invariant space interval*, S, defined by the expression,

$$S^2 = L^2 - c^2t^2 = L'^2 - c^2t'^2.$$

S will be real for spacelike relationships and will tend to zero as the relationship approaches the lightlike.

The concept of proper time interval is of particular interest as applied to the motion of a material body. In the course of time, a body changes its location in space so that specific values of its space coordinates belong to each value of t, as shown in Fig. 12. (Only two spatial dimensions, x and y, appear in the drawing; the cone of directions of propagation of a light signal is shown near the bottom.) The motion of the particle may be thought of as a series of "events," each with a well-defined location at a definite instant in time, $t_1, t_2, \ldots$, to be recorded by some observer.

These events must all be timelike relative to one another; the slope of the particle's trajectory in Fig.12 must be closer to that of the time axis than the slope of any of the light signals along the light cone. If any two of them were spacelike relative to one another, there would exist some other observer for whom these two events would be simultaneous; this observer would find one and the same particle at two distinct locations at once. Such an observation has never been made; if it should be made, one might question whether the two locations pertained to the same particle.

As all events along the trajectory of a particle are timelike with respect to each other, one can define a parameter along the trajectory, the particle's *proper time,* which increases as time goes on at such a rate that, for two closely adjacent instants in time, the increase in the particle's proper time equals the proper time interval, T, between these two events in the particle's history.

Proper time increases more slowly than coordinate time of any observer who is not at rest relative to the body; if observer and body are moving together, both increase at the same rate.

A good watch kept aboard the moving body will record proper time, and it will do so even when the body is accelerated. That, briefly, is the basis of the so-called twin paradox. Consider two twins, one of whom remains in a state of unaccelerated motion while the other travels out and back, changing his direction and hence accelerating at least once. When he eventually rejoins his less adventurous brother, the mercurial twin will be found to be the younger of the two; less proper time will have elapsed in his case. For proof consider an inertial frame in which the unaccelerated twin is at rest. His proper time will pass at the same rate as the coordinate time of that frame, whereas the proper time of the traveling twin will pass more slowly. The two twins are not interchangeable, as there is no inertial frame in which the traveling twin is always at rest.

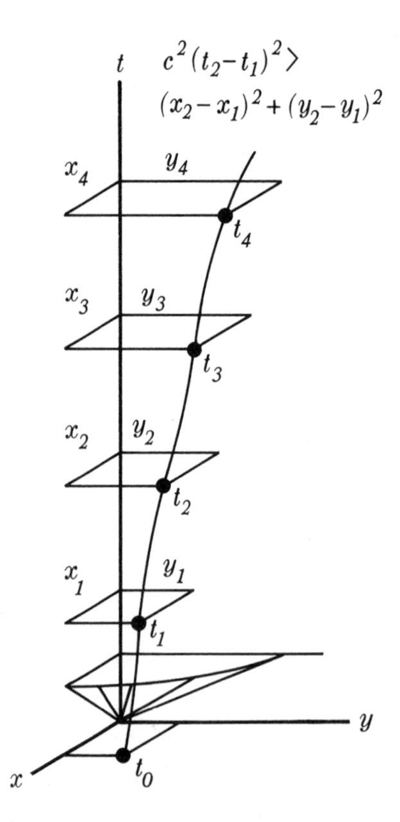

Fig. 12. Trajectory of a Particle

5 Minkowski's Four-dimensional World

Before the advent of relativity, space and time had been thought of as two entirely distinct domains, one having three dimensions, the other, one. This approach was entirely justified, because time data were presumed to be independent of any spatial relationship. Most of the observational information available to experimentalists up to the last decades of the nineteenth century involved motions so slow compared to the speed of light that the latter might almost as well have been infinite. The chief exceptions to this rule were observations on the eclipses of satellites of Jupiter, which led the Danish astronomer Olaus Roemer (1644–1710) to the discovery of the finite speed of propagation of light. By noting that the periodic occurrence of these eclipses was delayed or advanced, respectively, when earth and Jupiter were on opposite sides of the sun, or on the same side, Roemer concluded that it took light not quite twenty minutes to traverse a distance equal to the diameter of the earth's orbit about the sun. Combining this time of travel with the then best estimates of the distance of the earth from the sun, Roemer arrived at an estimate of the speed of light that was less than 20 per cent off the value accepted today on the strength of elaborate experiments.

Roemer's discovery is almost as old as Newton's development of a comprehensive theory of mechanics. For the next two centuries, the timing of astronomical observations was corrected for the finite time that it takes light to travel from the site of the event to the terrestrial observer, but there was no need to worry whether, for instance, the speed of light should be reckoned relative to the source, relative to the earth, or relative to some third frame of reference, such as the sun. For the speed of light is so large compared to all velocities encountered within the solar system (it is approximately 10,000 times the speed of the earth about the sun) that any small ambiguity or error in the correction for the transit time of light was negligible compared to other sources of error. Certainly Roemer's discovery gave no cause to question the possibility of unambiguously assigning to each event anywhere in the universe a time at which it took place. Time was considered universal, independent of the state of motion of the observer and independent of any particular frame of reference.

Spatial relations never were regarded as entirely independent of time. Consider two events, one of which takes place above New York, the other above Washington; the latter one hour after the other. From the point of view of a terrestrial observer, the sites of the two events are some two hundred miles apart. But suppose that the two events are two leaks appearing in the fuel line of an airplane engine. The engineer wants to find out whether these two leaks are independent or related failures, and he measures very precisely the distance between them, finding it to be a few inches. In terms of a frame of reference attached to the plane, the two events (the springing of the first leak and the springing of the second leak) are in close proximity to each other. The two statements, one that the two events were many miles apart, and the other, that their distance from each other amounts to a few inches, are not contradictory; each is correct, with respect to an appropriately chosen frame of reference. But the two statements about distance are compatible with each other only because they refer to two events that took place some time apart. Before the advent of relativity, it was universally agreed that the distance between two simultaneous events could be fixed unambiguously without regard to a particular frame of reference. It was also thought that the distance between two physical objects at rest relative to one another was well defined without the choice of a frame of reference.

Relativity destroyed the universality of time and space determinations and replaced it by the universality of the invariant interval. The *invariant interval* represents for any two events that are separated in space and in time a numerical relationship whose value is the same for all frames of reference. Hermann Minkowski (1864–1909), the Russian-born mathematician, suggested in 1908 that space and time were no longer useful as separate continua and should be replaced by a single four-dimensional continuum, space-time. In this fused continuum, the invariant interval would play a role similar to that of the ordinary distance in three-dimensional space.

In making this proposal, Minkowski did not intend to change the special theory of relativity, then barely three years old. Rather, he interpreted the theory from a new point of view, which has proved its value during the ensuing years. Formally, one can always combine two continua (called *manifolds* by mathematicians; the term implies the contiguity of the points composing the continuum) into one, by considering the combination of a point of the first manifold and of a point of the second manifold a point in the new ("product") manifold. Figure 13 illustrates the combination of a straight line, *L*, and a plane, *P*, into a new three-

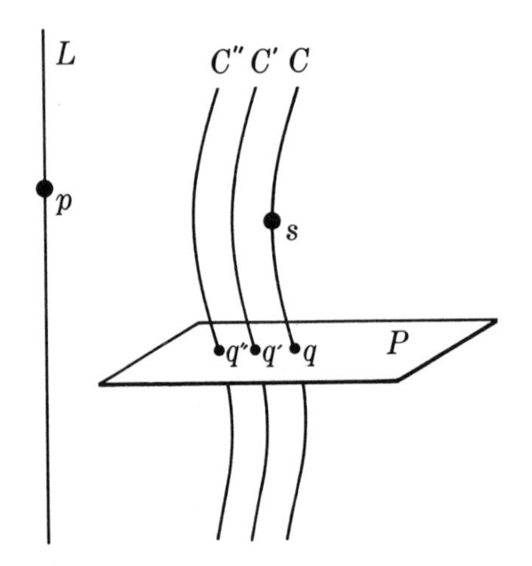

Fig. 13. "Product" of a Plane and of a Straight Line

dimensional space. This combination is made as follows: Choose a point p on L and a point q on P. Now pass through q a curve C, as similar curves C', C'', . . . , are passed through other points in the plane, q', q'' . . . . This construction is to be performed in such a manner that the curves C, C', C'', . . . do not intersect each other, but fill a three-dimensional region without gaps and holes. For instance, one could choose as the curves C, C', C'', . . . , the straight lines passing through the points of the plane P which are perpendicular to that plane; but other families of curves are equally acceptable. On each curve project the line L so that to each point of C there is assigned exactly one point of L, and vice versa. Let the point corresponding to p on L be designated by s. The point s is to "represent" the combination of p on L and of q on P. The totality of all points that "represent" combinations of pairs of points, always one on L and one on P, will then fill a three-dimensional region, and this region will be called the "product space" of L and P.

Figure 14 shows the "product" of two ordinary circles, C_1 and

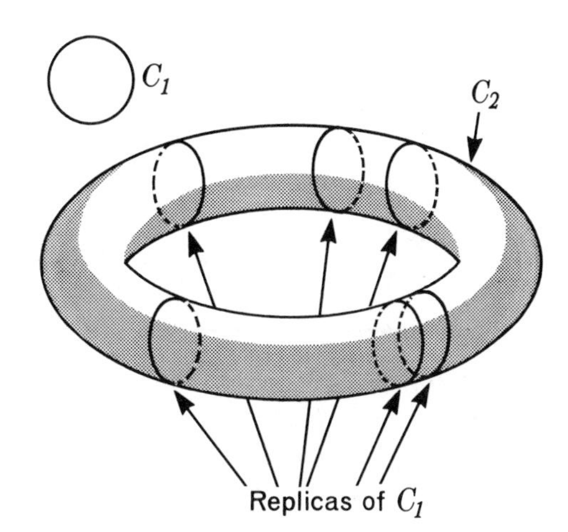

Fig. 14. The Torus as the "Product" of Two Circles

C_2. The product space is the surface of a doughnut, known as a *torus.*

Figure 15 shows the "product" of one direction in space (marked X) and the time axis, T. The combination of a point on X and a point of T is an *event,* an elementary occurrence that takes place at a particular location and at a definite time, iden-

Fig. 15. Two-dimensional "Event" Space

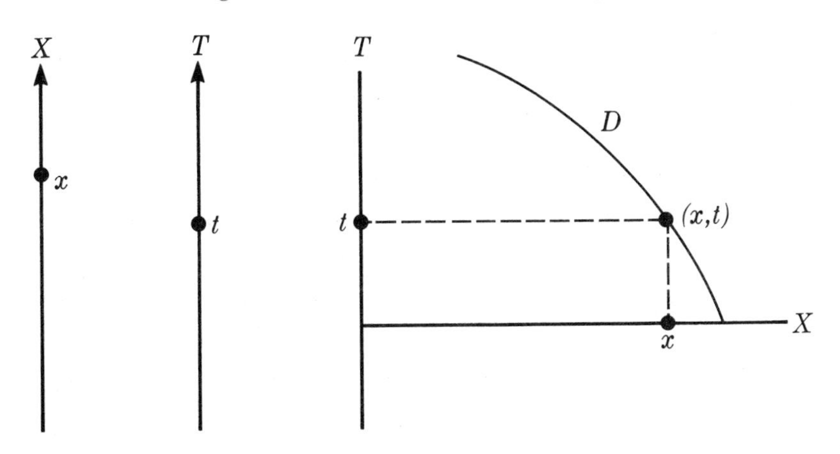

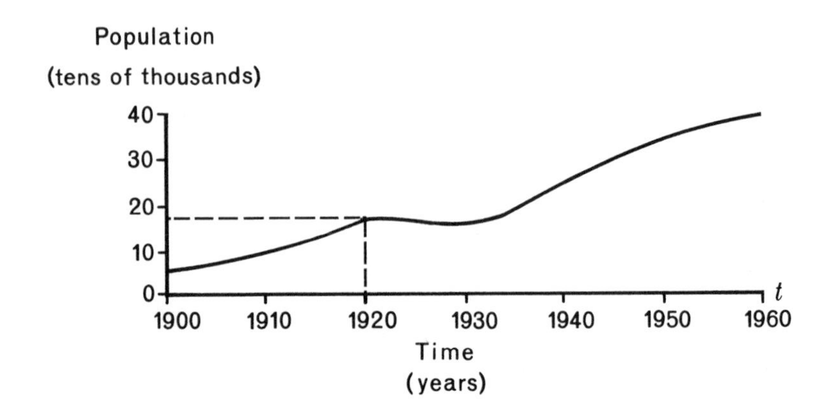

Fig. 16. Growth of a Population with Time

tified by the symbols x and t, respectively. In this sketch, the plane of the book page serves as the "product" of the two axes X and T. The point (x,t) is a single event. An infinite number of similar events, instants in the history of a freely falling massive body, form a curve, D.

Diagrams like Fig. 16 are commonly used to illustrate all kinds of relationships. By interpreting the coordinate X, not as a direction in space but as the number of people living in a certain city, one can diagram population growth as a new kind of space, the "product space" of population and time. A curve in the diagram represents population growth in that community. Drawing population movement as a continuous curve implies frequent counts, of course, or even continuous information.

In prerelativistic physics a diagram plotting spatial location of a body as a function of time would have been considered as just that, not as a point of departure for a new concept. Such a diagram would be useful for visualizing the relationship between the data on the same event collected by two observers, with the second moving, relative to the first, along the x axis at the rate v. (Both observers agree on the time to be assigned to the event, $t = t'$.) Figure 17 illustrates the relationship between x and x' by showing the relations connecting the coordinates (x, t) of an event with the coordinates (x', t') of the same event relative to a frame of reference moving to the right, relative to the first frame, at the rate v. At time t, the relative displacement amounts to a distance equal to vt. (This and the following diagrams show one space dimension rather than three in order to eliminate unessential details.)

According to the theory of relativity, this sketch must be modified, as shown in Fig. 18. Whereas in Fig. 17 the horizontal lines, which connect events that are taking place at the same time are common to all observers, this is not so in Fig. 18. The x axis and the x' axis are not parallel to each other because simultaneity of two events is a concept that is observer-dependent. The two events A and B of Fig. 19, which are simultaneous to one observer, take place at different times for the other observer, with B the earlier.

Figure 20 shows events that are in spacelike, lightlike, and

Fig. 17. Space-Time Transformation according to Classical (Newtonian) Physics

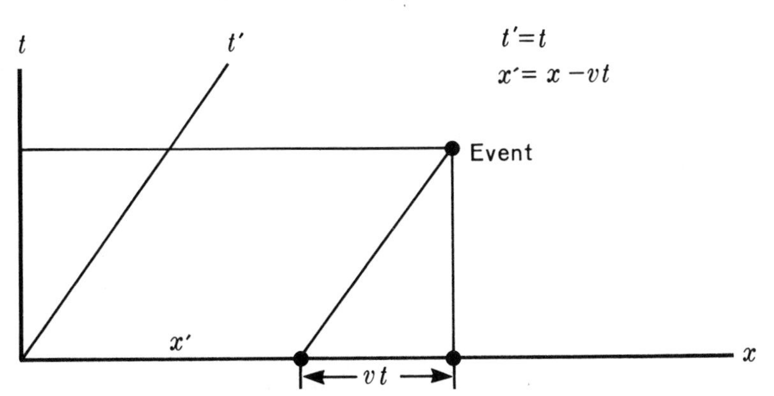

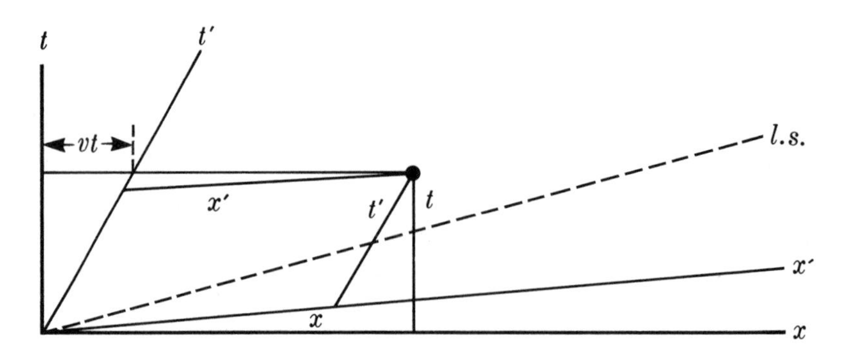

Fig. 18. Two Inertial Frames according to Einstein (Lorentz Transformation)

timelike relationships to each other. (The lines marked l. s. show the slopes at which light signals travel to right and left, respectively.) *A* and *B* are spacelike to each other; *C* and *D* and again *E* and *F* are in lightlike relations to each other; and *G* and *H* are timelike relative to each other.

To view space and time not as separate continua but as directions in a joint continuum or manifold, the space-time, the world, or the Minkowski universe—the three expressions are interchangeable and are all found in the literature—is justified be-

Fig. 19. Relativity of Simultaneity

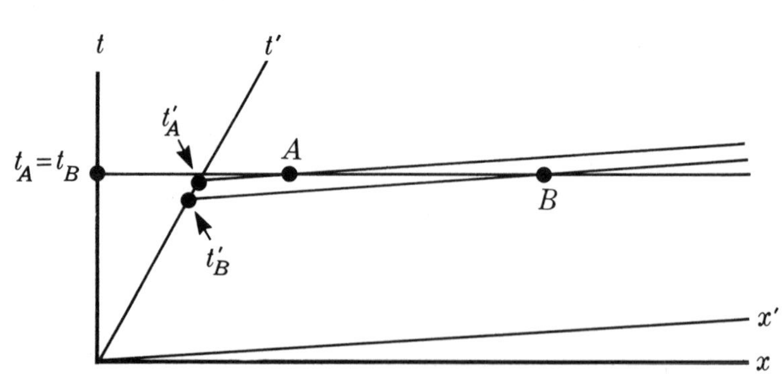

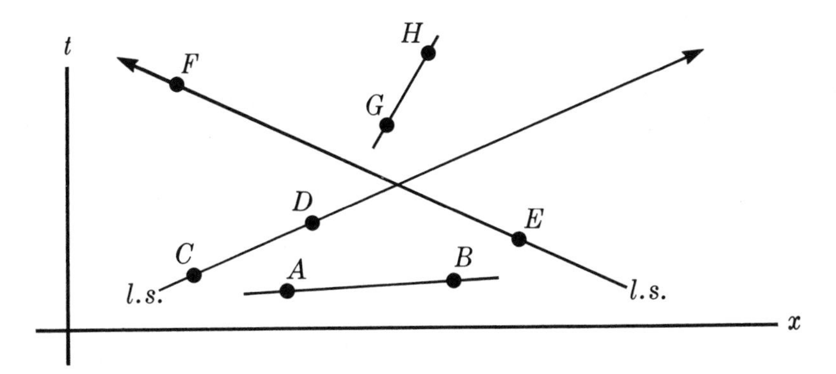

Fig. 20. Spacelike, Lightlike, and Timelike Relationships between Events

cause, in the transition from one inertial frame of reference to another, none of these axes remains unchanged, in contrast to the situation sketched in Fig. 17.

Minkowski's approach is useful because it suggests the need to search for concepts that are "natural" in four-dimensional space-time and independent of any decomposition into separate space and time.

In any space-time diagram, the *history* of a particle (its movements in the course of time) is represented by a curve. From Minkowski's point of view, this curve is the only natural representation of that history, and one whose essential features are independent of the chosen inertial frame of reference. It is called the *world curve* of the particle. This curve, shown in Fig. 12, must be timelike in all of its parts; hence, in any inertial frame of reference, the particle cannot reach or exceed the speed of light.

In one respect Minkowski's geometry differs fundamentally from any ordinary Euclidean geometry. In ordinary geometry the fundamental relationship between any two points is the distance of one from the other. This distance is always positive. In Minkowski's geometry two points may be timelike to each other, spacelike, or lightlike. Accordingly, the square of the invariant interval between these two points may be positive, negative, or zero. The distance of Euclid has this in common with Minkowski's invariant interval: the values of both are independent of the choice of coordinate system (frame of reference, in Minkowski's case) and represent intrinsic geometric relationships.

The world points of the Minkowski universe are to be described in terms of a coordinate system which possesses four coordinate axes. One axis corresponds to the ordinary time scale; the other three, to the coordinate axes in ordinary space. In ordinary space, the preferred coordinate systems are called *Cartesian* coordinates, after the French mathematician René Descartes (1596–1650). In a Cartesian coordinate system, all three coordinate axes are straight, and at right angles to each other. In Minkowski geometry there are also preferred (four-dimensional) coordinate systems, which correspond to an observer in unaccelerated motion who uses Cartesian coordinates for spatial fixation and ordinary clocks for timing events. These coordinate systems are named for Lorentz; they are called *Lorentz coordinate systems*, or *Lorentz frames* (of reference). The formulas for space-time intervals to be found in the preceding chapter are based on the use of Lorentz frames.

In ordinary geometry, transitions from one to another Cartesian coordinate system, x and y, to another, x' and y' are of several different kinds. One type, called a *translation*, involves a displacement of the *origin* of the coordinate system (the point at which all coordinate values are set zero, marked 0 or $0'$), without changes in the directions of the coordinate axes; this type of transition is illustrated in Fig. 21, where a and b indicate the amount of displacement. In this and the following two figures only two spatial coordinates are shown; in each of these figures P designates

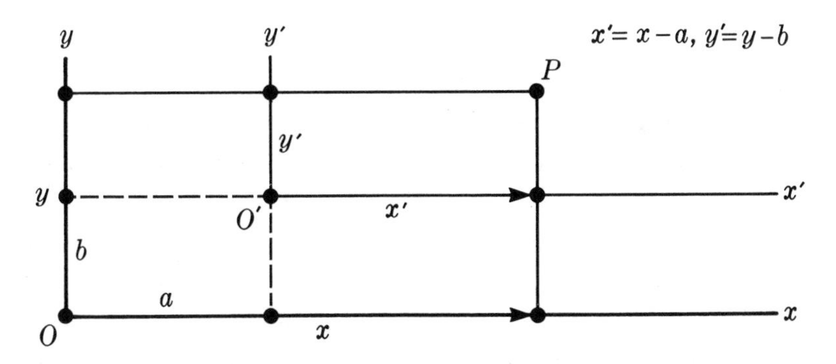

Fig. 21. Coordinate Translation

the point whose coordinates are shown with respect to two different Cartesian coordinate systems. The second type of transition to a new coordinate system consists of a *rotation* of the coordinate axes about the origin, which remains unchanged, as illustrated in Fig. 22. The angle of rotation is marked θ. Combining a translation with a rotation results in the general type of *coordinate transformation* (transition from one coordinate system to another) that leads from one Cartesian coordinate system to

Fig. 22. Coordinate Rotation

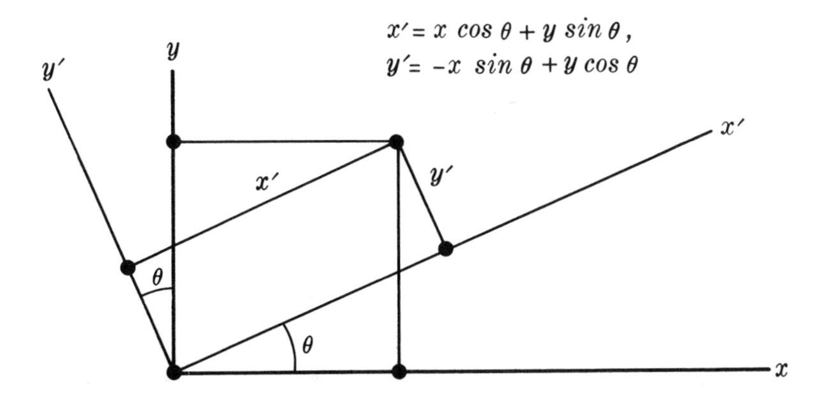

another (Fig. 23). If both the initial and the final coordinate systems are Cartesian, one speaks of an *orthogonal transformation.*

In the Minkowski universe, the transition from one inertial frame of reference to another, or from one Lorentz frame to another, is called a *Lorentz transformation.* The most general Lorentz transformations involve a translation of the origin of the time axis, a translation of the origin of the space coordinates, a change in the directions of the space coordinates, and a change in the (unaccelerated) state of motion of the observer as in Figs. 11 and 18. A quantity, or a relationship, that remains unchanged when a Lorentz transformation is performed is called *Lorentz-invariant.* Mathematically, the physicist's search for laws that are the same for all observers is a search for Lorentz-invariant relationships.

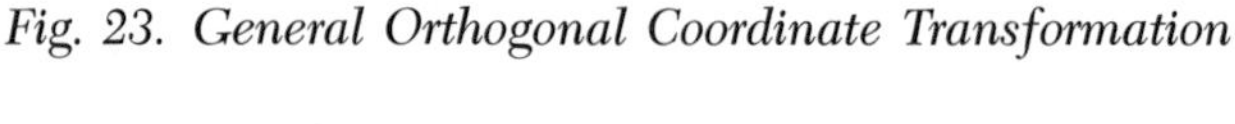

Fig. 23. General Orthogonal Coordinate Transformation

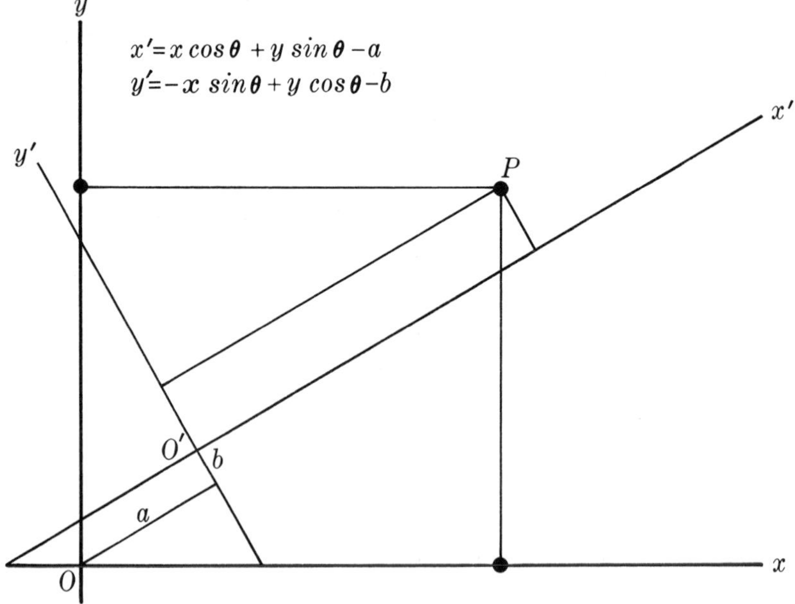

In the subsequent history of the theory of relativity and of its application to all the special areas of physics, this mathematical approach has amply demonstrated its heuristic power. Today's physics can scarcely be imagined without Minkowski's four-dimensional point of view. Nevertheless, all discussion of the theory of relativity may be cast in a form in which (for a given observer) the three dimensions of space are separated from the time axis. Einstein's early papers were, of course, written in just that language. These papers (1905–1908) already contain the relativistic theory of electricity and magnetism, and a good deal of relativistic mechanics. Since spatial and time measurements call for quite different apparatus, separation of space and time is a prerequisite to the design of any experimental or observational procedure. The three-plus-one and the four-dimensional languages in relativity tend to complement each other.

At every world point of the Minkowski universe, the totality of all directions in space and time forms a pencil of arrows pointing away from that point. Some of these directions are lightlike, some timelike, and others spacelike. The lightlike directions form two cones, which are called the *light cones* associated with that world point. (Fig. 24: two spatial, x and y, axes and a time axis, t, are shown in this diagram.) Each of the two light cones has its own interior, which is formed entirely of timelike directions. The directions on one light cone, and in its interior, point into the future, and this whole light cone is called the *future light cone;* the other light cone is referred to as the *past light cone.*

A Lorentz transformation changes a space-time direction into some other direction, but it will not change its basic character. It will always turn a timelike direction into another timelike direction; similarly, the characteristics of lightlike and of spacelike directions are preserved intact. As all lightlike directions are turned into lightlike directions, each of the two light cones is

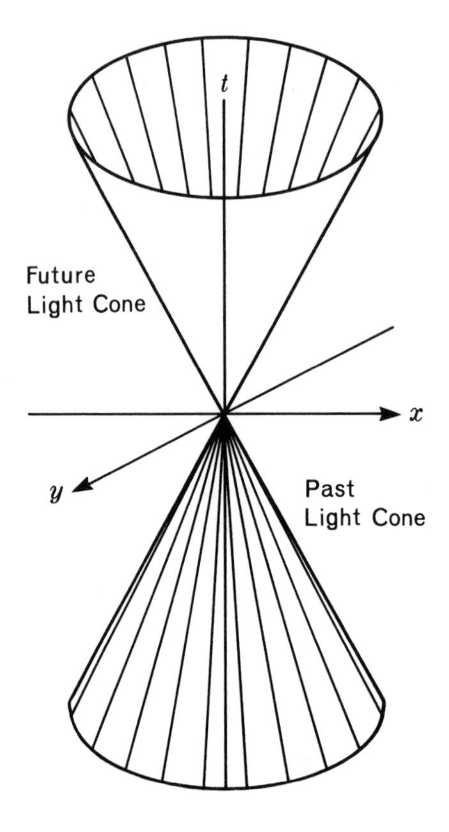

Fig. 24. The Two Light Cones

kept intact; each lightlike direction is shifted about only on its own light cone. These cones of directions, which do not depend on the choice of frame of reference, possess no analogs in ordinary Euclidean space. They exist in Minkowski geometry only because the invariant interval, which in other respects resembles the distance between points in Euclidean spaces, is zero between certain pairs of points, and the direction pointing from one point of such a pair to the other is "special," lightlike.

Light cones incorporate all those directions along which light pulses may propagate. In particular, the past light cone contains all those directions in space-time from which light borne information may reach an observer at the location and at the time identified with the fixed world point on which the cone is centered. Thus each direction on the past light cone may be identified with a point on the *celestial sphere,* which represents our view of the sky and the stars, just as a globe represents our view of the earth. The angle that an astronomer measures between two star images is the angle between two light directions on the past light cone. If two observers move relative to each other at considerable speed, their respective determinations of the angle between two given stars will not agree. Indeed relative locations of stars depend noticeably on the motion of the earth relative to a fixed inertial frame of reference, and this velocity changes in the course of half a year by two ten-thousandths (2×10^{-4}) of the speed of light on account of the earth's revolution about the sun. This effect is known as *aberration;* it was discovered by the English astronomer James Bradley (1693–1762) in 1725 and amounts to an apparent displacement of up to two-thirds of a minute of arc. Each star completes its apparent elliptical circuit in the course of one year.

6 Mass, Energy, Momentum

Relativity was born of electromagnetic theory. But as the electromagnetic field manifests itself through its interaction with electrically charged particles, relativity was bound to come to grips with the behavior of particles and to assimilate the notions of Newton's mechanics into its own body of concepts.

According to Newton's laws, two interacting bodies exert on each other forces that are equal in magnitude (and possess opposite directions). As a result they produce changes in one another's respective velocities that are inverse to their respective masses. Suppose now that one of the two interacting particles has a very large velocity, close to the speed of light. According to relativity, no body can travel at a speed greater than that of light; some built-in circumstance must prevent any velocity from being boosted above that limit. This is one reason that Newtonian mechanics requires some modification in order to be compatible with relativistic requirements.

Further, it is basic to Newtonian mechanics that if two bodies interact with each other, their respective velocities change so that their common center of mass (often called *center of gravity* or *centroid*) continues to move at the same velocity. Thus the law of inertia, that a body will not change its velocity in the absence of external forces, holds not only for elementary particles but also for macroscopic bodies, which are composed of many constituent particles exerting forces on each other. If this law is to remain valid in relativistic physics, the motion of the center of mass if unaccelerated in one Lorentz frame must have a constant velocity in all Lorentz frames.

Because velocities transform differently under Lorentz transformations than in Newtonian physics, the common center of mass will not conform automatically to the requirement of unaccelerated motion in all Lorentz frames. In fact, Einstein discovered that, to achieve this objective, he had to assume that the mass of each constituent particle depended on its velocity, and hence was not the same in all Lorentz frames. The faster a particle moved, the greater its mass. The mass of a moving particle was found to exceed its mass at rest by an amount equal to its kinetic energy, divided by the square of c. This new dependence of the mass on the velocity assured that no particle could ever be accelerated to a speed exceeding the speed of light. It provided the first inkling of the awesome equivalence of mass and energy, $E = mc^2$, the formula that has been confirmed in subsequent history far beyond the original theoretical prediction. Elementary particles, the ultimate constituents of matter, can be converted to radiant energy, directly or through some intermediate stages. The small *mass defects,* by which the masses of atomic nuclei deviate from the masses of their constituents, represent energy, the so-called binding energy of an atomic nucleus. and these energies may be liberated in the course of nuclear reactions, gradually in nuclear reactors, and explosively in nuclear arms.

The mass of a body measured in a Lorentz frame in which it is at rest is called its *rest mass.* The rest mass is an intrinsic property of a physical body, whereas the mass that governs its behavior in interactions with other bodies, its *relativistic mass,* depends on the relative motion of object and observer as well. The sum of the relativistic masses of several interacting bodies remains unchanged through the interaction, but the rest masses can change. For instance, an atomic nucleus will undergo radioactive decay into several smaller structures and in the process also emit a *gamma ray* (a very energetic electromagnetic pulse of a wavelength much shorter than either visible light or even ordinary X rays). The rest masses of the end products will add to less than the rest mass of the original nucleus, but the sum of the relativistic masses, including the relativistic mass of the gamma ray, will equal that of the nucleus undergoing radioactive decay.

A quantity which remains unchanged in the course of the history of a physical system that is isolated from its surroundings is said to be *conserved.* In Newtonian physics, both mass and energy are such conserved quantities. Any law that asserts that some quantity, such as mass or energy, remains constant in the course of time is called a *conservation law.*

In relativistic physics, energy and (relativistic) mass are equivalent, except that they are measured in different units. The unit of mass is larger by a factor c^2 than the unit of energy. Hence, a relatively modest amount of mass in a nuclear bomb (less than one part in a hundred of the active material in a thermonuclear device) represents an enormous amount of explosive energy once it is set free.

A quantity that is conserved in the case of an isolated physical system may change in a system that is interacting with its surroundings. The rate of change in a given domain of space is determined by the amount of flux of the conserved quantity across the bounding surface. For instance, the amount of (relativistic) mass in a given volume will increase if additional mass enters that volume and decrease if there is a flux of mass directed outward. If the emphasis is placed on the balancing of the rate of change of a conserved quantity by a compensating flux, rather than on its conservation in the absence of flux, one speaks of a *law of continuity.* Ordinarily, laws of continuity are formulated in terms of the *density* of the conserved quantity, for instance, *mass density* (mass per unit volume). The density at a point in space increases at the rate at which the corresponding flux *converges,* or decreases as the flux diverges. Only converging and diverging fluxes matter, as a uniform flux will bring into a region as much of the conserved quantity on one side as it carries out on the other. The two terms *conservation law* and *law of continuity* refer to two different formulations of the same kind of law, which asserts that certain physical quantities cannot be freely created or destroyed, but merely displaced.

Before the theory of relativity was formulated, the energy of a body was conceived of as its ability to perform work on other physical objects, and this ability was considered to reside in both its location in space and its state of motion. A heavy weight on top of a tall building can drive an engine in the process of being lowered to the ground, if, by means of a rope, the weight turns a winch as it drops to the ground. This part of the energy is called its *potential energy.* As the weight reaches the ground, it loses its ability to perform further work; hence its potential energy is reduced. That part of the energy of a body residing in its state of motion is called its *kinetic energy.* An automobile traveling on the highway at fifty miles an hour will continue to be able to perform work even when its engine is shut down and it is coasting. If it hits a telegraph pole, it will cause the pole to move, and in the process the car itself will slow down; in doing work, the automobile loses part of its kinetic energy.

Under appropriate circumstances, a body can perform work by having part or all of its rest mass converted into available energy, without involving any potential or kinetic energy that it may possess. For this reason, it makes sense to consider the body's energy to be composed of potential, kinetic, and *rest energy,* the latter being c^2 times its rest mass. The sum total of all forms of energy of a physical object is called its *total* (*relativistic*) *energy.* The total energy equals c^2 times the relativistic mass of the object.

Although for every inertial observer the relativistic mass (and with it the total energy) is conserved, relativistic mass and energy do not have equal values for two observers moving differently, because these two observers will find different values for the velocity of the same object. In this respect, relativistic mechanics resembles Newtonian mechanics. According to classical mechanics, the total energy (kinetic plus potential energy) of an object, or of a system of particles, is conserved in any particular inertial frame of reference. But whereas potential energy is independent of the frame of reference used, kinetic energy has different values in different frames of reference, and so does total energy. Both in classical and in relativistic mechanics, then, the total energy is conserved; its value remains unchanged in the course of time, for any one choice of inertial frame of reference; but though constant in change, the value of the energy will change on transition from one inertial frame of reference to another.

Is there any law by which the energy in one frame of reference is related to the energy in another frame? In both Newtonian and relativistic mechanics, the answer is affirmative. Between the two values of the energy in the two frames of reference, there is a mathematical relation which involves not only the relative velocity between the two frames, but also another quantity, the *linear momentum.* The linear momentum of a particle is the product of its mass and its velocity. As such it represents a measure of the impact that the particle will make on colliding with a stationary object. The linear momentum is a *vector,* a quantity that possesses direction—the direction of the particle's velocity—and which, in a Cartesian coordinate system, is conveniently decomposed into its three components parallel to the three coordinate axes, just like the velocity.

The linear momentum of a system of particles is also defined; it is the *vector sum* of the linear momenta of the constituent particles. The sum of several vectors is defined not as just the arithmetical sum of their magnitudes (every vector has not only direction but a magnitude as well) but as the vector that one would obtain by laying the constituent vectors end to end with proper regard for their directions. In terms of components, one obtains the components of a vector sum by adding the appropriate components of the constituents. For instance, the x component of the sum is the (arithmetical) sum of the x components of all the constituents (*terms*). The vector sum of the linear momenta of several particles equals the product of their total mass and of the velocity of the common center of mass.

Given the conservation of mass, Newton's law of inertia may also be stated in the form that the linear momentum of an isolated system does not change in time. It remains constant, in spite of internal interactions, because the changes in linear momentum that any two constituents bring about in each other will cancel each other out, in accordance with the principle that the forces causing these changes are equal and opposite. Again, there is a law of continuity, this time controlling the linear momentum: The linear momentum present in any region of space can change only as a flux of linear momentum is permitted to pass through the bounding surface of that region. In relativistic mechanics, the correct expression for the linear momentum, the one that leads to a conserved quantity, is obtained by multiplying the velocity of an object by its relativistic mass. The linear momentum in one Lorentz frame is related to the value of the linear momentum in another Lorentz frame by a law that also involves the relativistic mass, and hence the energy. As a result, in order to calculate the values of total energy (or relativistic mass) or linear momentum in one Lorentz frame of reference, knowledge of all these quantities in another Lorentz frame is both necessary and sufficient. The numerical values of four quantities—the energy and the three components of the linear momentum—in one Lorentz frame determine the numerical values of the corresponding quantities in any other Lorentz frame.

Further mathematical analysis shows that these four quantities behave in the four-dimensional Minkowski universe as the four components of a vector, a geometric object that, in the four-dimensional continuum, has both magnitude and direction and may be decomposed into its four components parallel to the four space-and-time axes of any chosen Lorentz coordinate system.

In ordinary three-dimensional space, the three components a, b, c of a vector combine to yield its magnitude V by the equation

$$V^2 = a^2 + b^2 + c^2$$

according to Pythagoras' formula. Given two different Cartesian coordinate systems, the components of the vector will differ in the two systems, but the sum of their squares will be the same; in mathematical language, the magnitude V is *invariant* with respect to orthogonal coordinate transformations (orthogonal transformations, remember, are those coordinate transformations that lead from one Cartesian coordinate system to another). Likewise, if the relativistic energy of a body be denoted by the symbol E, and the components of its linear momentum with respect to the spatial

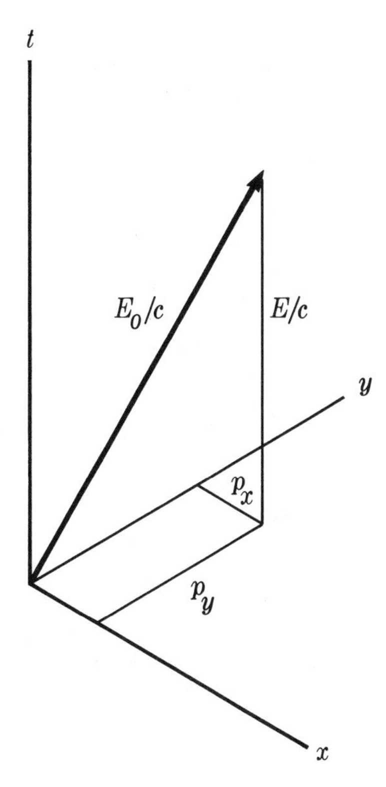

Fig. 25. The Energy-Momentum Vector in Minkowski Space-Time

x, y, and z axes by p_x, p_y, and p_z, respectively, then the *rest energy,* E_0, is given by the expression

$$c^{-2}\,E_0^2 = c^{-2}E^2 - p_x^2 - p_y^2 - p_z^2$$

The quantities E/c, p_x, p_y, and p_z may thus be considered the components of a vector in Minkowski space-time (Fig. 25; only two spatial dimensions are shown here and in Fig. 26). The components of this vector, usually called the *energy-momentum vector,* take different numerical values in different Lorentz frames, but its magnitude is the same in all Lorentz frames; it is invariant with respect to Lorentz transformations, or Lorentz-invariant.

The whole of the energy-momentum vector is conserved; the

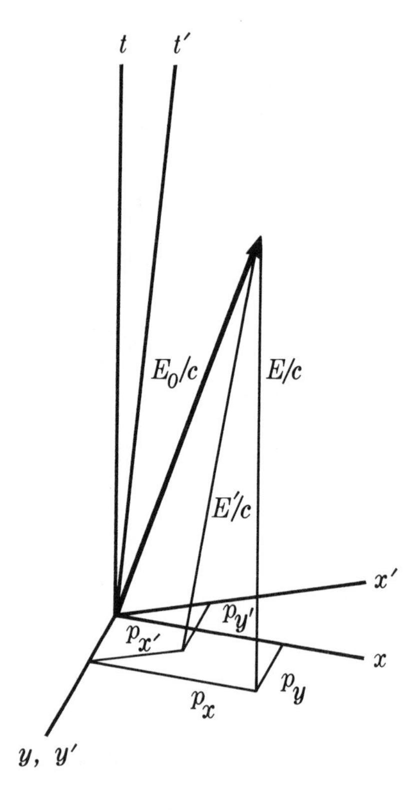

Fig. 26. Components of the Energy-Momentum Vector in Two Different Lorentz Frames

energy-momentum vector belonging to an isolated mechanical system will not change its magnitude or its direction in Minkowski space-time in the course of the history of that system, no matter what its internal evolution. Conservation of the whole vector implies conservation of its separate four components in any one Lorentz frame, even though their values will differ from one Lorentz frame to the next.

For later presentation of the general theory of relativity, it is necessary to elaborate the discussion of energy and momentum, and of their conservation laws, in terms of their respective densities, and thus to arrive at the appropriate laws of continuity. To present this technique properly, it will be helpful to consider another example, the electric charge, which lacks some of the complexities of the energy-momentum vector.

The electric charge of a physical system is a *scalar* (a single quantity whose value is the same in all coordinate systems). The electric charge of an isolated physical system is conserved. It cannot change in the course of time as long as no charge is brought into or out of the system. The total charge of a system equals the (arithmetic) sum of the charges of the constituents.

Instead of speaking of the total charge of the system one may introduce the notion of *charge density,* the amount of charge per unit of volume. As the volume of a region of space is different in different Lorentz frames, because of the relativity of distances and lengths, the charge density is not the same for all observers. The charge density turns out to be one component of a four-dimensional vector, whose remaining three components represent the flux of charge, the *electric current density.* Electric current density is defined as the amount of charge that crosses an area of unit size, 1 cm^2, for example, per unit time, where the unit area is to be perpendicular to the direction of the flow of charge. In terms of ordinary three-dimensional space, the current density is a vector, and its three components represent the flux in the x direction across an area in the (y, z) plane, the flux in the y direction, and in the z direction.

The law of conservation of charge may now be stated in terms of the electric charge density and of the electric current density: The rate of increase of the electric charge density equals the convergence of electric current density, the rate at which current flows into a region in excess of current carrying charge out. This *law of continuity of electric charge* is equivalent to its conservation law, because, if there were no conservation, the rate of increase of charge density might be either greater or smaller than accounted for by flux; charge might be generated out of nothing, or might disappear altogether.

Returning to the four-dimensional language, the rate of change of the electric charge density in the course of time may be interpreted as but one component of the flux into a four-dimensional domain, which, when added to the other three components, adds up to zero. Four-dimensionally, then, the law of continuity of electric charge states that the excess of flow of charge into a four-dimensional region over the outgo from that same region vanishes. If one could represent the four-dimensional flow pattern graphically by a system of *streamlines*, curves running in the direction of the flow, whose closeness to each other at each point represents the magnitude of the flow there, then, according to the law of continuity, all the streamlines would be curves without beginning or end, running either in closed loops or from infinity to infinity.

The analogous formulation of the laws of conservation of energy (mass) and linear momentum requires introduction of the notions of energy density, energy flux density, linear momentum density, and linear momentum flux density. Whereas total energy and total linear momentum of a system form a world vector, with four components, the densities form a structure with up to sixteen components. This structure might have been called a vector of the second order, but usually it is called a *tensor.* The number of components is somewhat reduced from the maximum number of sixteen, because the energy flux density is the same as the linear momentum density; the linear momentum flux density, also known as the *stress,* has the further property that the flux of x-linear momentum in the y direction equals the flux of y-linear momentum in the x direction, and so forth. The final number of components of the *energy-stress tensor* is ten.

7 Flat Space—Curved Space

For describing geometric relations in ordinary space, Cartesian coordinate systems are preferred, because most geometric relations assume their simplest form in terms of Cartesian coordinates; but this preference is a matter of convenience, and there are exceptions. In Newtonian mechanics, for instance, the orbits of the planets about the sun are much more conveniently described in a coordinate system adapted to the sun's preeminent role in this system, and to the uniformity of its gravitational effects in all directions. The spherical coordinates of a point P consist of its distance (r) from the coordinate origin (the sun) and of the two angles (θ and ϕ) that specify the direction in which one must go from the sun in order to reach that point (Fig. 27). One among these three coordinates, the distance from the sun, determines the strength of gravitational attraction, whereas, in Cartesian coordinates (x, y, z in Fig. 27) all three coordinates are involved evenly. Spherical coordinates provide an example of a curvilinear coordinate system, in which some or all of the coordinate axes are curves rather than straight lines. Likewise, in the Minkowski universe there are rare occasions in which some other coordinate system is preferable to a Lorentz frame.

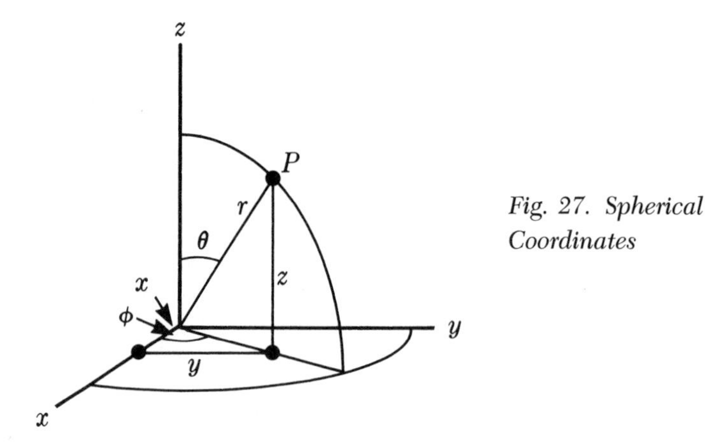

Fig. 27. Spherical Coordinates

Lorentz frames are the analogs of the Cartesian coordinate systems of ordinary geometry. Their axes, both space and time, are all straight lines, and they are at right angles to each other. But curvilinear coordinate systems are imaginable in Minkowski space-time, and they are used on occasion. For instance, one may wish to describe the universe as it looks to an observer who is undergoing a constant acceleration, as on a space rocket while the rocket is being fired. Figure 28 shows the trajectory of such an accelerated observer, with the coordinates x and t. According to Newtonian physics, his velocity would increase without bound, as indicated in Fig. 28a (the $[x, t]$ diagram is a parabola). According to relativity, the observer's velocity will increase, but it never can exceed the speed of light, *l.s.*; the resulting curve is shown in Fig. 28b. It is a hyperbola whose slope is asymptotically limited by the light cone. For simplicity, those two spatial axes not involved in the motion have been omitted. Figure 29 shows a space-time coordinate system, x, x', t, appropriate to such an observer. Peculiarly, his distance from the point O never changes because this distance is to be determined along a line perpendicular to the direction of flight of the observer at any given instant and

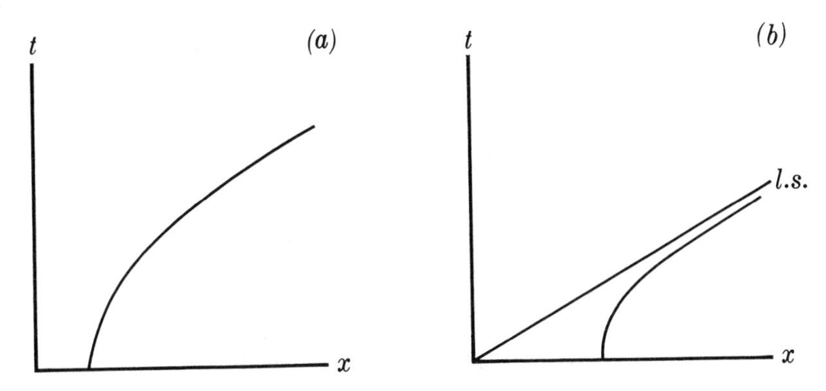

Fig. 28. Trajectories of a Uniformly Accelerated Observer

representing "the same time" as determined by that observer. Thus, the "lengths" of the lines AO, BO, and CO of Fig. 29 are all equal.

Cartesian and Lorentz coordinate systems are *straight-line coordinate systems;* spherical coordinate systems and accelerated frames of reference in the four-dimensional Minkowski universe

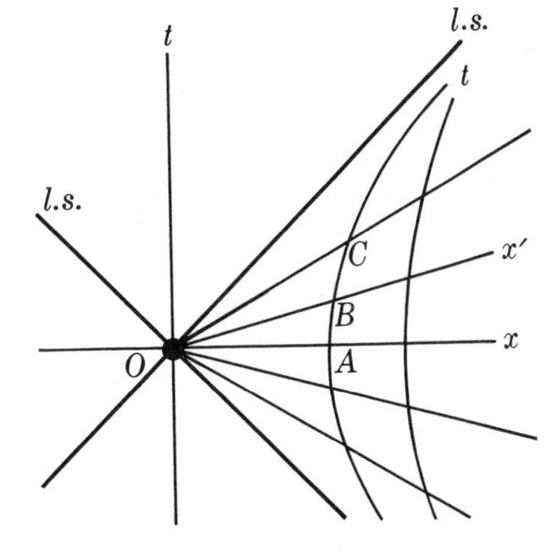

Fig. 29. Coordinate System Appropriate to the Observer of Fig. 28b

are *curvilinear coordinate systems.* Whether one performs geometric constructions in one kind of coordinate system or the other affects, of course, the mathematical form of the constructs; the description of physical situations will look different in different kinds of coordinate systems. But the choice of coordinate system is a choice of the mode of description; it does not affect the intrinsic properties of the continuum to be described. Independently of the manner of description, spaces and continua possess intrinsic geometric properties; one of these, to be described now, is called the *curvature* of a space or continuum. (Confusingly, a space will be called *curved* unless it is *flat,* and this property is quite distinct from the choice of a curvilinear coordinate system.)

A space (or manifold) is called *curved* if it cannot be covered by a coordinate system that may be properly called a straight-line coordinate system. The space is called *flat* if a straight-line coordinate system exists, whether it is used or not. For the purposes of this definition, a *straight line,* or more properly a *geodesic,* is that curve between any two given points that represents the shortest connecting path between them. A coordinate system is straight-line if its axes (including their parallels) everywhere are straight lines; in that case, the axes of two given kinds (say, the x axes and the y axes, again including their parallels throughout the space) intersect at the same angle everywhere. No such coordinate system can be constructed in a curved space. All continua presented up to this point have been flat.

An example of a curved manifold, in this case a two-dimensional manifold, is provided by the surface of a sphere. On a spherical surface, the geodesics are the so-called great circles, which indeed have many properties ordinarily associated with straight lines (Fig. 30). On the surface of the earth, the meridians and the equator are great circles. Because a great circle passing through two given points is the shortest curve connecting them (shortest,

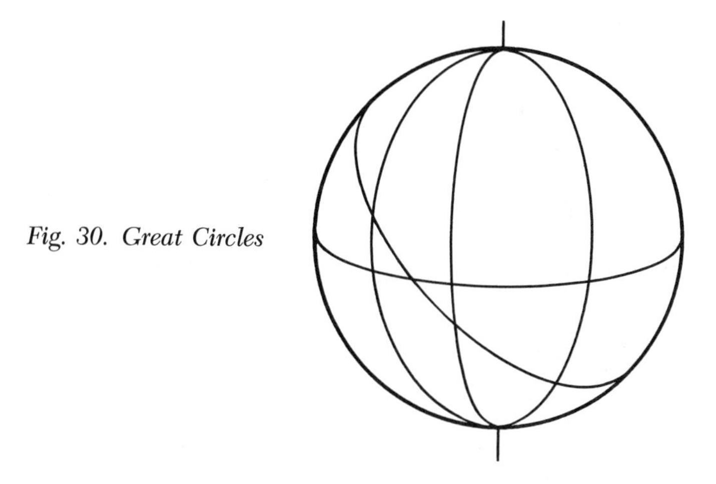

Fig. 30. Great Circles

that is, on the spherical surface), long-distance air routes are frequently chosen to be great-circle routes. Great circles, and segments of great circles, may be used for constructing geometric figures. Thus, one speaks of spherical triangles or quadrilaterals (Fig. 31). But there are no spherical parallelograms. If one attempts

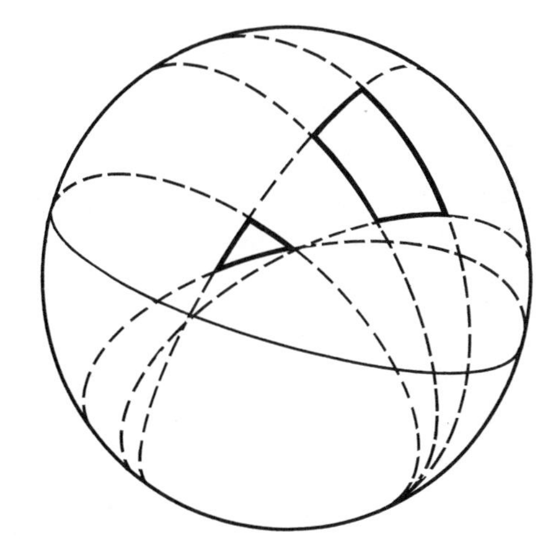

Fig. 31. Spherical Triangle; Spherical Quadrilateral

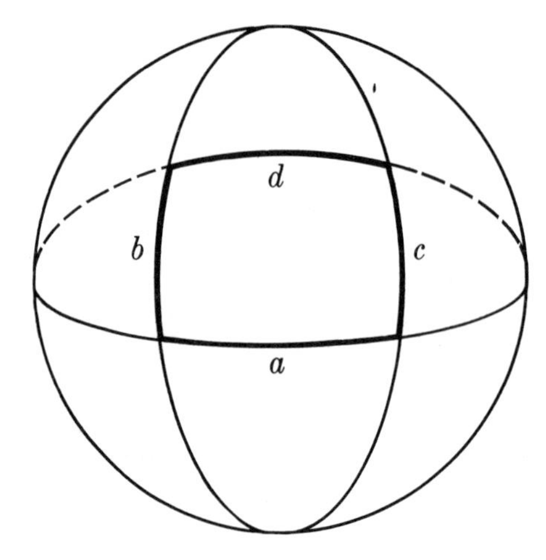

Fig. 32. Attempt to Construct a Rectangle on the Spherical Surface

to construct one (Fig. 32) by marking off from the end points of some great-circle segment, *a*, two new sides, *b* and *c*, of equal length and perpendicular to *a*, then the fourth side, *d*, of the quadrilateral will be shorter than the initial segment, *a*. One can set out with a piece of the equator and erect on its end points two meridian segments of equal length, whose far end points will then come to lie on a parallel circle. That the distance between these two points is smaller than the original piece of equator is obvious. If one makes the meridian segments extend to the pole, the two end points will coincide, and the result of the construction will be not a quadrilateral but a triangle, with two right angles (Fig. 33). Hence, the sum of the three angles of this triangle will exceed 180° by whatever the third angle amounts to, whereas in a flat two-dimensional space (a plane), the three angles of a triangle always add up to precisely 180°.

The notion of *parallel transport* offers a good approach to understanding the characteristic properties of curved spaces. Given two points in a space and a vector at one of them, one may construct that vector at the other point which is parallel to the first.

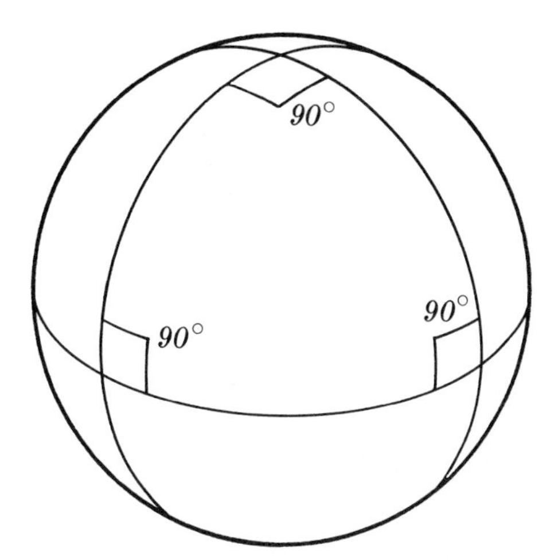

Fig. 33. Spherical Triangle in which the Sum of the Angles Equals 270°

To this end, one connects the two points by a geodesic and then transports the original vector along the geodesic, seeing to it that the angle between the straight line and the vector remains constant in the course of the transfer, that the vector is not rotated about the straight-line track while sliding along it, and that the vector does not change its magnitude. The same procedure will accomplish parallel transport of a vector along a course consisting of several straight-line segments (Fig. 34). Many paths are available along which the parallel transfer from one point to another may be managed. In a flat space, the end result depends not on the path of transfer but solely on the original vector; a single direction is parallel at each point of space to that originally given elsewhere. In a curved space, the result of parallel transfer depends on both the original vector and the path of transport chosen. If vector v_0 (Fig. 35) is transported parallel to B and thence to D the result is the vector v_1, whereas parallel transport of v_0 from A to D via C results in vector v_2. Parallel transport of a vector along a broken straight-line path that ultimately leads back to the point of departure will result in a new vector at the original

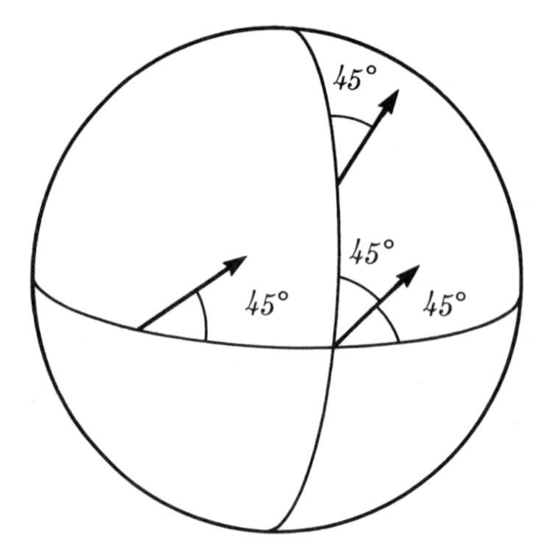

Fig. 34. Parallel Transport of a Vector on the Surface of a Sphere

point; the new vector differs from the original, even though the proper procedure for parallel transport has been employed at each segment of the loop.

To return to the previous example of the surface of a sphere, a spherical triangle that consists of two meridians and a segment

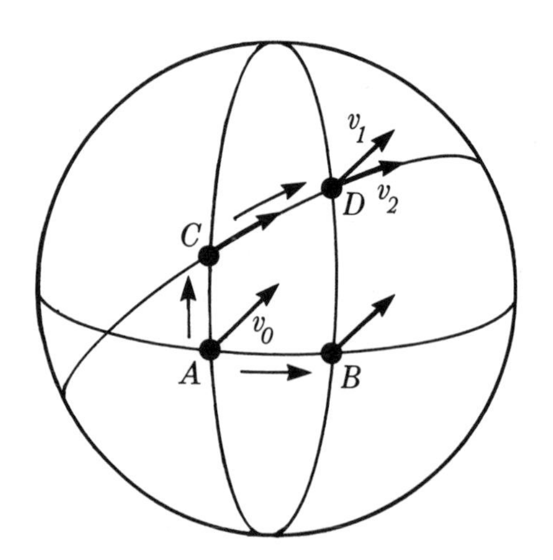

Fig. 35. Parallel Transport along Alternative Paths

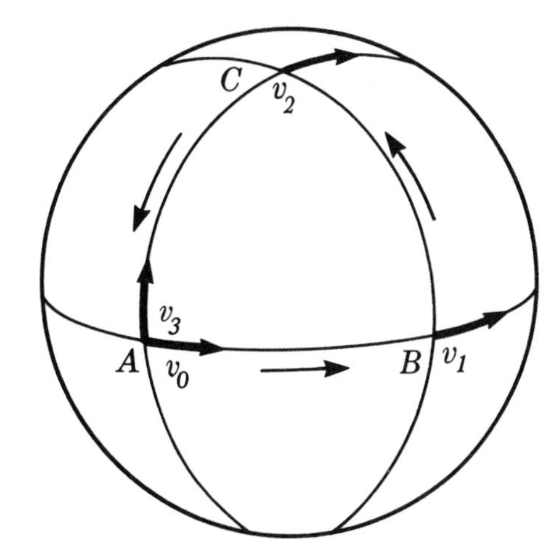

Fig. 36. Parallel Transport along a Closed Path

of the equator is a closed loop consisting of segments each of which is geodesic. If the segment of the equator spans 90° in geographical longitude, as in Fig. 33, it is not difficult to follow the fate of a vector that is transported parallel once around that loop (Fig. 36). Suppose the circuit is begun at the point on the equator farthest west, *A*, and that the vector v_0 is pointing due east. Parallel transport along the equator to *B* results in a vector v_1 still pointing east; during the trip along the eastern meridian up to the pole, *C*, the parallel-transported vector will still point east until it arrives at the pole. At the pole, the vector v_2 will be at right angles to the eastern meridian, and hence tangential to the western meridian. On being slid south along the western meridian back to the equator and to *A* the vector will continue to be tangential to the meridian and to point north; and it will have that heading on reaching its point of departure, v_3, 90° away from its original heading.

The Minkowski universe is a flat space. To a direction established at one world point there corresponds exactly one parallel direction at every other world point. They all will be timelike, spacelike, or lightlike, depending on the character of the initial choice. Consider now that timelike direction, for a given Lorentz frame, which is parallel to the time axis. A particle whose world line has this direction is at rest relative to the given Lorentz frame. That this particular direction can be identified unambiguously with directions at other world points implies that the notion of rest in a given Lorentz frame is meaningful throughout space and time. Only because of the flatness of the Minkowski universe is the state of motion of two physical objects with respect to each other well defined regardless of the distance between them.

Probably, Einstein and Minkowski did not even make a deliberate choice when they constructed the space-time of the special theory of relativity to be flat. Usually, physicists become conscious of such assumptions in their theories only when these theories become questionable. So it was here. Within less than ten years of Minkowski's paper, Einstein was led to modify the geometry of the Minkowski universe by postulating that in the presence of gravitational fields space-time is curved.

II / GENERAL RELATIVITY

8 Relativity and Gravitation

The special theory of relativity emerged in response to the seeming inconsistencies between Newtonian physics and the new physics of the electromagnetic field. Reasserting the principle of relativity and denying the existence of absolute motion and absolute rest, Einstein achieved the reconciliation of what appeared to be mutually exclusive requirements—the universal speed of light versus the equivalence of all inertial frames of reference—by changing profoundly our ideas of space and time. More than in any other area of the natural sciences, any change in space and time concepts requires major adjustments in physics, because space and time are the scaffolding in terms of which physical processes must be described. Such adjustments are particularly called for in theories dealing with the fundamental aspects of particles and their interactions.

The need for revision was not equally urgent in all areas of science. The exploration of the secrets of nature has always proceeded on many different levels of profundity and complexity. A living organism, for instance, is a system of enormous complexity, whose dynamics cannot be reduced to detailed description of all the interactions of its constituent atoms and molecules. No one would seriously suggest that biological research be deferred until all the elementary processes which, taken together, make up life are fully comprehended separately. But when it comes to understanding the spectrum of the hydrogen atom, which consists of but two particles, one proton and one electron, the scientist cannot be satisfied until all the details of the spectrum are fully explained in terms of the law of interaction between these two particles.

Improved understanding at one level of complexity usually contributes to an understanding of adjacent levels but may be relatively unimportant at levels far removed. A better comprehension of the dynamics of atoms will facilitate the study of molecules and will often contribute to the physics of crystals; but its bearing on the taxonomy of reptiles will be negligible. Similarly, there are but few points of contact between the physics of atoms and the physics of elementary particles, though there is some mutual relevance. Nor did the conceptual revolution known as the special theory of relativity engulf all areas of physics equally.

For several reasons, relativity cannot be disregarded in the investigation of the ultimate constituents of matter and of the fundamental forces. In interactions between elementary particles and fields large velocities, which tend to emphasize the differences between Newtonian and relativistic physics, are most often encountered. By contrast, the molecules in a gas, the constituent atoms of a crystal, or those of a living organism generally do not possess velocities at which relativistic effects are important. In studying interactions at the fundamental level, the depth of comprehension desired—even more than the occurrence of extreme velocities—is a chief reason for paying heed to relativity.

The more deeply an investigation delves into ultimate constituents and the smaller the number of participating particles and forces, the more stringent become the demands to understand fully the functioning and structure of each component. For that reason, once Einstein and other physicists were persuaded that the theory of relativity of 1905 was going to supersede Newtonian physics, they turned to reexamine the fundamental properties of particles and force fields. The most important subject for such a reexamination was gravitation.

Gravitation was the touchstone of Newton's theory of physical interactions, and his mechanics achieved its most impressive successes in explaining the properties of planetary and satellite orbits in our solar system. Any new comprehensive theory of physics had to duplicate these achievements to merit acceptance. But Newtonian mechanics depends completely on the notion of universal time, in that the forces that the sun and the planets exert on each other are determined by their respective distances from each other *at the same time*. One refers to Newton's description of gravitational interaction as "action at a distance." As the distance between any two bodies is forever changing while both are traversing their respective trajectories, it is essential that the distance be defined unambiguously at every instant.

Whereas in Newtonian physics the distance between two physical objects at any particular instant is well defined whether they are at rest or not, the relativity of time and space data, and their dependence on the observer in relativity, give rise to ambiguities. Figure 37 illustrates these difficulties: Consider the trajectories of two objects A and B, in a space-time diagram, in which again, for the sake of simplicity, but one of the spatial dimensions is shown. A and B are in motion with respect to each other, but the trajectories of both are timelike, as is indicated by the slopes of the two lightlike directions, *l.s.*. The two sets of coordinates shown, x and t, and x' and t', correspond to two Lorentz frames; hence both are equally legitimate. Given some particular instant in the history of A, say, T, the distance of B from A is measured along two dif-

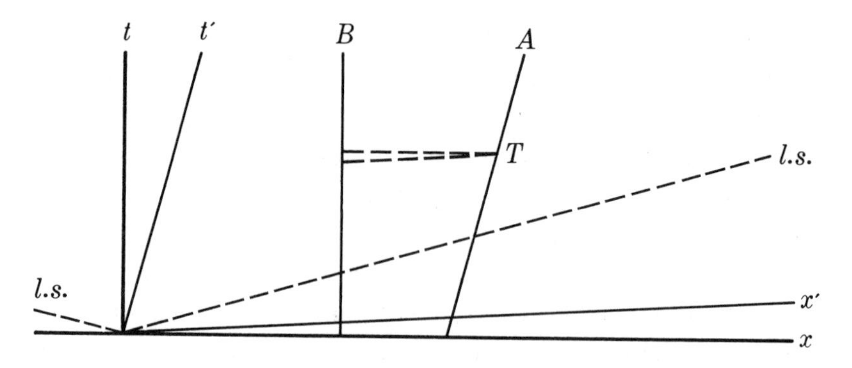

Fig. 37. Ambiguity of Distance

ferent connecting lines (shown dashed) in the two Lorentz frames, and the resulting values for the distance come out different. There is no way in which a particular Lorentz frame can be selected in order to define the "true" distance between a number of astronomical bodies, which would then be utilized in a computation of the gravitational interactions between them. True, in the solar system and even in our galaxy, relative motions of the constituents are so slow that it makes almost no difference whether one selects the galaxy as a whole as a Lorentz frame or one in which the center of mass of the solar system is at rest. But though the smallness of relativistic effects may explain why using Newtonian methods of calculation has led to highly satisfactory results, as a matter of principle the employment of mutually exclusive space-time concepts in the treatment of electromagnetic and of gravitational interactions cannot be accepted.

From one point of view, a somewhat similar situation arises in electromagnetics, as the force between two charges at rest is also determined by the distance between them; it obeys an inverse-square law, similar to Newton's law of gravitation. This law, known in electricity as Coulomb's law, is named for the French physicist Charles Augustin de Coulomb (1736–1806). Coulomb's law of electric forces and Newton's law of gravitational forces differ from one another in that in Coulomb's law, the electric charges play the role reserved in Newton's law for the masses. In electricity, the apparent contradiction between relativity and the law of electric forces is resolved by the discovery that Coulomb's law holds rigorously only if the charged bodies do not move with respect to each other. If they are at rest, the "appropriate" frame of reference to be used is the one frame in which all charges involved are at rest. Whenever the charged particles move relative to each other, Coulomb's law must be replaced by a much more complex interaction, which can be best described in terms of the fields generated by the interacting charges.

Every electric charge gives rise to an electric field, and if not at rest also to a magnetic field. Both fields spread at a finite rate, which is the speed of propagation of light. The force experienced by any one charge is in turn determined by the field generated at its location by all the other charges far enough back in time so as to arrive just then. This complex interplay between charges and field turns out to obey the same rules regardless of the state of motion of the observer who determines the pertinent data through his measurements. If the two charged particles are at rest relative to each other, the particle-field-particle interplay reduces to the simple law of Coulomb.

Why not resolve the inconsistency between the relativity of time and Newton's law of gravitation in the same fashion? It would be necessary to introduce the notion of a gravitational field, which would propagate somewhat like electric and magnetic fields and which would mediate the interaction between gravitating celestial bodies consistent with relativity. This gravitational interaction should reduce to Newton's law whenever the relative velocities of the astronomical bodies involved are small compared to the speed of light. Einstein did try to construct a relativistic theory of gravitation along these lines, but he was frustrated by one circumstance: Nothing was known about high-velocity gravitational interaction as such, but some information was available concerning the effect of high speeds on the sources of the gravitational field, the masses.

Large velocities affect masses differently from electric charges. Whereas a body's electric charge has the same value for all observers, its mass depends on its speed relative to the observer. The higher the speed, the larger the observed mass. For a given body, the smallest mass will be reported by an observer with respect to whom that body is at rest. This value is called the rest mass of the body. For all other observers, the mass appears larger by an amount proportional to its kinetic energy, as discussed in Chapter 6. Its magnitude would become infinite in a frame of reference in which the body is traveling at the speed of light. Of course, such a frame of reference can be only approximated. Because the magnitudes of the sources of gravitation depend so much on the frame of reference in which they are measured, the resulting field is bound to be more complex than the electromagnetic field. Einstein concluded that the gravitational field probably was a so-called tensor field, to be described by a greater number of components than those of the electromagnetic field.

As further guiding principles, Einstein postulated that the laws of the gravitational field should be obtainable by mathematical procedures closely analogous to those leading to the laws of electricity and magnetism, and that the resulting laws themselves should be similar in form. These postulates appeared reasonable because of the nearly identical forms of Newton's and Coulomb's laws.

Proceeding along these lines, Einstein found that he could construct several different theories that appeared to conform equally well to his requirements. These theories led to some effects that differed from theory to theory but those were so minute that in the first decade of the twentieth century no attempt to reach a decision by experimentation or observation could have been envisaged. Some other point of view was needed to lead without ambiguity to a relativistic theory of gravitation. Eventually, Einstein discovered this new point of view in the principle of equivalence (Chapter 1), according to which the acceleration does not depend on the characteristics of the body that is subjected to gravitational forces.

9 The Relativity of Free Fall

Both Newtonian physics and the special theory of relativity postulate the existence of inertial frames of reference. To repeat, an *inertial frame of reference* is a frame with respect to which bodies move uniformly and without acceleration whenever there are no external forces. The experimental determination of such an inertial frame depends on being able to remove some bodies (test bodies, as it were) from all external forces and to confirm by experimental procedures that such forces are indeed absent. How does one bring about the forcefree state? The usual answer is that the test bodies must be removed sufficiently far from all other bodies that might be the sources of forces, as all forces known will wane and tend to zero at sufficiently large distances between source and object.

Thus, a survey of the surroundings of a proposed test object will reveal whether there are any potential sources of external forces left. But such a survey depends on the availability of optical or other instruments which permit the scanning of large regions of space. One would like to know whether there are any procedures available for ascertaining the absence of forces that are local in character and that are based more directly on the examination of the behavior of the test bodies themselves. Such procedures are indeed feasible to an extent, if one employs a variety of test bodies with different characteristics. Suppose, for instance, that one wishes to confirm the absence of an ambient electric field. If one has available several test bodies, some charged positively, some negatively, some uncharged, no electric field will be present if it is discovered that all these different test bodies are accelerated in the same manner with respect to an arbitrarily chosen (not necessarily inertial) frame of reference. If an electric field were present, the positively charged bodies should experience electric forces directed opposite to those suffered by the negatively charged bodies. Analogous procedures might be developed to test for other types of force fields.

Such an approach is unavailable when one seeks to eliminate gravitational fields. Because of the proportionality of inertial and gravitational masses of all possible physical objects, any variety of test particles will undergo identical accelerations in any ambient gravitational field, whether this field be weak or strong. How does one tell, then, that there is a gravitational field, such as the one caused by the large mass of the sun? Kepler and Newton provided the answer: observe the acceleration of all the planets, including the earth, toward the sun (Fig. 38). The individual planets accelerate relative to each other, as well as relative to the distant fixed stars, because at any one time they are at different distances and at different directions from the sun. The differences in their accelerations correspond to differences in the ambient gravitational field at their respective locations. Two celestial bodies relatively close to one another, as a planet and its satellites, fall toward the sun together; most of the difference in their respective accelerations is accounted for by the gravitational attractions they exert on each

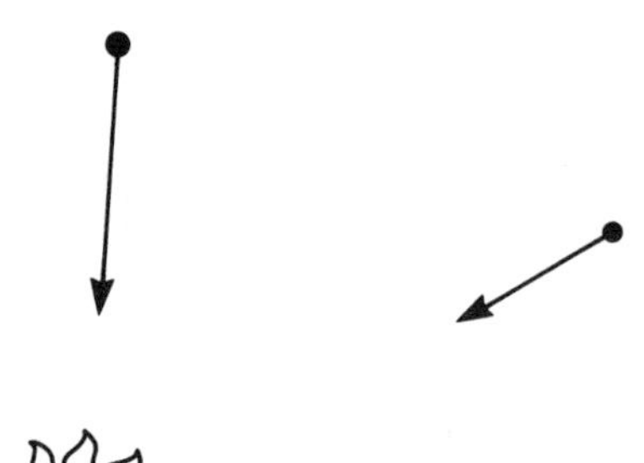

Fig. 38. Gravitational Accelerations Experienced by Two Planets

other. The vectors a and c in Fig. 39 represent respectively the accelerations that the planet and moon would experience if each were present without the other; d and e the accelerations resulting from their mutual attraction if the sun were eliminated; and b and f their actual accelerations in the presence of all three bodies.

To sum up, whereas the presence of an electric (or any other) force field can be detected by the differential effects that such a field exerts on different test particles even at the same location, all test particles brought into the same gravitational field will exhibit the same acceleration. By local experiment, which does not draw on observation of distant stars, the presence of a gravitational field can be established only by its inhomogeneities, which lead to differences in the accelerations of test particles at different locations within that field.

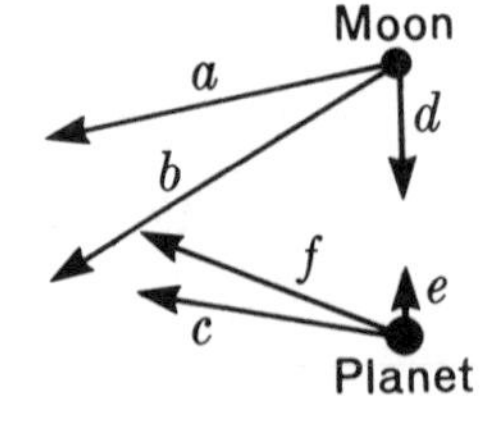

Fig. 39. Planet and Moon

In the absence of gravitational fields, inertial frames of reference can be constructed by means of experiments performed locally, provided that the experimenter has available various test particles whose diverse physical properties are known. In the presence of gravitational fields, all test particles undergo the same gravitational acceleration; hence this acceleration becomes known only through comparison with objects at diverse locations, not through comparison of diverse test particles at the same location. For this reason astronomers have constructed the inertial frames of reference that they need to study the dynamics of the solar system, or of the galaxy, by referring to external objects, such as distant galaxies.

Even in the presence of gravitational fields there is a class of frames of reference that can be obtained from purely local experimentation. Because all the gravitational accelerations at a given location are the same, in magnitude as well as in direction, these accelerations will be zero with respect to a frame of reference that is accelerated in tune with the physical objects that are subject to gravity but to no other forces. Such a frame of reference is called a *free-falling frame of reference.* A free-falling frame of reference differs, however, from all frames of reference previously discussed in that it cannot be extended arbitrarily far through space and time. It is well defined only in the neighborhood of a world point, in a limited region of space and for a limited period of time. Because of the inhomogeneity of all gravitational fields, any attempt to extend a free-falling frame of reference to far distances or to long periods of time must fail through internal inconsistencies. If the frame is extended by rigid rods, gravitational accelerations relative to the rods will be noticeable at sufficient distances from the region where the frame is truly free-falling. If all points of the frame are made to fall freely, the distances between them will change in the course of time.

A frame that cannot be extended might be called a *local frame of reference.* Free-falling frames of reference are local in this sense. With respect to a free-falling frame of reference, material bodies will be unaccelerated if they are free from non-gravitational forces.

Free-falling frames of reference are identical with inertial frames of reference when no gravitational fields are present, and in that case they are, of course, extensible. Their extensibility is lost when there are gravitational fields. That free-falling frames of reference exist at all, though merely as local frames of reference, is a result of the principle of equivalence obeyed by all gravitational effects. This same principle is responsible for the breakdown of local procedures to construct inertial frames of reference when gravitational fields are present.

Einstein accepted the principle of equivalence as a fundamental property of gravitation. He came to recognize that the concept of the infinitely extensible inertial frame of reference might have to be abandoned in favor of the local free-falling frame of reference; only by taking this step would the principle of equivalence be made part of the foundations of physics. But through this approach physicists might gain a deeper insight into the nature of gravitation. The presence of a gravitational field would be tantamount to the non-extensibility of the local free-falling frames of reference; hence, study of gravitational fields must be focused less on the local strength of that field (which would always be zero with respect to the free-falling frame of reference) than on its inhomogeneities. The validity of this approach, which in effect denies the universal existence of inertial frames of reference, depends on showing that there is no compelling reason to take inertial frames for granted, even though they were accepted for several hundred years.

In order to extend an inertial frame of reference throughout space and time, one must be able to compare distant clocks with each other and to extend straight rods of specified length deep into the regions to be surveyed. To determine whether clocks are in fact synchronized, and whether rods are indeed straight, one ultimately depends on light signals and other electromagnetic radiation as the universally available and reliable carriers of information. In empty space, light propagates at the uniform speed c and along straight paths. Will it continue to do so in the presence of gravitational fields?

Light carries energy; hence it possesses mass. In the presence of gravitation it must undergo acceleration. To be sure, acceleration of a particle depends on its present state of motion. If the acceleration is in the direction of its velocity, acceleration at high velocities must be small, as no particle can ever exceed the speed of light. According to the laws of the special theory of relativity, the forward acceleration of any object that already has achieved the speed of light is zero. But if the force of gravity acts at right angles to the direction of flight of a particle, then its trajectory should be curved. The amount of curvature can be estimated on the strength of the assumption, and extension of the principle of equivalence, that the light path will appear straight to a free-falling observer. What matters in the context of the present argument is that if light is bent by gravitation, then the normal procedure for constructing an extended inertial frame of reference will not work.

Can this procedure be refined so as to compensate for the gravitational bending of light? If the principle of equivalence is to be trusted, no correction can be determined experimentally by comparing the curved path of light with some straight trajectory. Any signal traveling at the speed c in the same direction will follow a like trajectory; any particle traveling at a lesser speed will follow a trajectory curved even more strongly. Accordingly, the necessary corrections can at best be made on the strength of a comprehensive theory of gravitation; the construction of an inertial frame of reference must follow, rather than precede, the elaboration of that theory.

Actual observations of the bending of light rays by the gravitational field of the sun have been performed by means of a stratagem that is based on the apparent immobility of distant fixed stars. During a total eclipse of the sun, it is possible to take pictures of the field of stars surrounding the darkened location of the sun, because, during its occultation, the light emanating from the sun does not interfere with the visibility of fainter objects. If the pictures taken during the solar eclipse are compared with pictures of the same region of the heavens taken at night, those stars closest to the limb of the sun during the eclipse are found to be displaced slightly, by amounts that are inversely proportional to the distance of the stellar image from the sun. The displacement of the star images directly at the sun's limb is 1.75 seconds of arc. These observations have been performed, with varying success, during almost every total solar eclipse since 1919. When weather conditions have been favorable, the observations have agreed well with the predictions of the general theory of relativity.

Though before Einstein's conjecture the existence of inertial frames of reference had been taken for granted, non-inertial frames had frequently been used for convenience. Such a frame of reference might, for instance, be tied to a rotating platform. Persons who have never in their lives tried to balance on a rotating disk of the sort sometimes shown at county fairs are nevertheless quite used to living on the rotating platform that is represented by our earth. Any frame of reference that is tied to the walls of our terrestrial laboratory is non-inertial in that it participates in the rotation of the earth about its axis. As a result of this rotation, the plane of a pendulum turns slowly at a rate that depends on the geographical latitude of the laboratory. A pendulum that shows this rotation is called a *Foucault pendulum.* (There is a Foucault pendulum on exhibit in the lobby of the Assembly Building of the United Nations in New York.) Another effect to be observed in a rotating frame of reference is that objects initially at rest in that frame are accelerated away from the axis of rotation; this acceleration is known as *centrifugal acceleration.*

Accelerations that are observed merely because of the choice of a non-inertial frame of reference are generally known as *inertial accelerations.* Inertial accelerations are independent of the physical characteristics of the test bodies on which the observations are carried out, as they merely reflect the degree to which the chosen frame of reference fails to be inertial, and depend at most on the initial state of motion of each test body. In this respect, inertial accelerations resemble gravitational accelerations, which, according to the principle of equivalence, are also common to all free-falling bodies in the same location. In a free-falling frame of reference neither inertial nor gravitational accelerations are observable.

On occasion the sequence of this argument has been turned around, and the principle of equivalence has been proclaimed in the form that, by local experiment, gravitational and inertial accelerations are indistinguishable. The equality of gravitational and inertial mass then follows as a deduction.

A space capsule falls freely whenever all rocket engines are turned off, and a frame of reference connected with it is a free-falling frame. Hence the astronauts inside will have no sensation of gravity even though the capsule is well within the reach of earth's gravity field. Generally, a free-falling frame of reference in which the presence of a gravitational field is ignored will be indistinguishable from an inertial frame of reference when gravitational fields are truly absent, except that the latter can be extended, whereas the former is purely local. Conversely, the presence of a gravitational field leads to effects in an inertial frame of reference that are indistinguishable, locally, from inertial accelerations as they would be observed in a non-inertial frame. Hence, the existence of inertial frames in our universe, in which gravitating masses are present, would have to be inferred from the development of experimental procedures that rely on the availability of very visible but distant unaccelerated objects, procedures which moreover compensate for the bending of light rays, which is unavoidable in the presence of large masses.

In all probability, no such experimental program is feasible in principle. Procedures intended to fix an inertial frame of reference in an indefinitely extended region of space and for an indefinite length of time involve ambiguities that cannot be resolved but which lead to uncertainties so minute that in practical astronomy they are overshadowed by ordinary observational errors.

Einstein proposed forgoing all attempts to reinstate the role of inertial frames of reference. Locally, their role would have to be taken over by the free-falling frames of reference, which, however, are not extensible. Among extended frames of reference, there would be no hierarchy, no criteria for selecting a class of special or superior frames. All frames of reference were to be accepted as equally valid.

10 The Principle of General Covariance

In the language of four-dimensional geometry, a frame of reference is a *(four-dimensional) coordinate system.* According to Einstein, the geometry of space-time does not lend itself to the selection of a special class of coordinate systems for formulating the laws of nature, but all reasonably continuous and smooth coordinate systems will serve equally well. This formal principle is known as the *principle of general covariance.* Its effect is to require that laws of nature take the same form in all coordinate systems. In particular, it must be impossible to classify coordinate systems into those in which the axes are straight lines and those in which the axes are curved. If gravitational fields are completely absent, straight-line coordinate systems—inertial frames of reference—exist, but this is an exceptional situation. Typically, gravitational fields are all-pervasive, though very weak in intergalactic space. There can be no curves in space-time that have all the properties ordinarily associated with straight lines.

The purpose of all coordinate systems, whether straight-line or curved, is to identify world points, events well localized in space and time, by means of sets of four numbers. If no straight-line coordinate systems exist, the four-dimensional space-time continuum must be *curved.* Its curvature is the geometric counterpart of gravitational fields, which are always inhomogeneous. In the absence of gravitational fields, space-time is flat, and Minkowski's original construction is valid. But if gravitational fields are present, Minkowski's geometry is too rigid; it must be modified so as to include the possibility of curved manifolds. (The term *manifold* is used here to denote a four-dimensional space-time continuum which is not necessarily flat.)

That space-time is curved does not in any way contradict the notion of the free-falling frame of reference. In geometric terms, such a free-falling frame represents a local coordinate system, defined only in the immediate vicinity of a given world point, whose coordinate axes are as nearly straight-line and mutually perpendicular as is consistent with the over-all curvature of space-time.

Gravitational fields are always inhomogeneous; the strength of the field, and the resulting accelerations, vary from point to point; for this reason, a free-falling frame of reference cannot be extended indefinitely. If one constructs a free-falling frame of reference near the surface of the earth, such a frame can be extended into a fairly large region before any difficulties arise. When, however, the frame is extended into a region so large that the inhomogeneities of the gravitational field become noticeable, then, at far points, the frame is either no longer free-falling or not straight-line. These distances might amount to several hundred miles, depending on the degree of accuracy required. (A more precise estimate is given in the next chapter.)

If space-time is curved, then there are no frames of reference that qualify as being inertial everywhere, no rectilinear coordinate systems. Free-falling frames of reference are coordinate systems whose axes are geodesics only in the vicinity of a point. If extended beyond that region they have no properties that would distinguish them from other curvilinear coordinate systems; in a curved manifold any reasonably smooth (though curvilinear) coordinate system is as suitable as any other.

Curved manifolds have certain geometric properties which have nothing to do with the choice of coordinate system. Space-time manifolds possess at each point a double cone of lightlike directions, half of which point into the future, half into the past. The *curvature* of the manifold itself is defined in terms of the change of direction that a vector suffers when transported parallel about a small closed loop. The change in direction depends on the original direction of the vector and also on the orientation of the two-dimensional surface in which the loop is embedded; given this orientation, the change in direction is proportional to the area enclosed by the loop. Hence the magnitude of the curvature of a manifold is expressible in terms of the change in direction (in degrees of arc) per unit area circumnavigated.

All these and other properties of manifolds can be described in terms of any chosen coordinate system. The form of this description varies, for obvious reasons, from coordinate system to coordinate system. Coordinate systems differ from one another in that a given space-time point carries different numerical values of the coordinates in different systems. For instance each particular component of the curvature may vary from point to point. (A *component of the curvature* is defined by the combination of the original direction of a vector, its change in another direction, and the orientation of the surface in which the loop lies. In four dimensions there are twenty independent components.) The description of that variation will depend on the particular identification of each point (the choice of coordinate system), as well as on the direction of a vector and on the orientation of the loop.

The descriptions of direction and orientation depend on the directions chosen for the coordinate axes.

Whereas the description of a particular manifold and its geometry will vary with the choice of coordinate system, some possible relations between geometric parameters do, if satisfied, take the same form in every coordinate system. Such relations between geometric properties are said to be *covariant.* Manifolds that satisfy any such relation at all points form a special class, distinguishable (by this relationship) from all other manifolds. Flat manifolds are a case in point. Flat manifolds are a special class of manifolds in which all components of the curvature vanish throughout the manifold. The vanishing of all components of the curvature can be described by certain mathematical equations that take the same form in all coordinate systems; flat manifolds also, incidentally, admit of a special class of coordinate systems, rectilinear systems. Though it is usually advantageous to avail oneself of this opportunity and to choose rectilinear coordinates, the description of a manifold as flat does not depend on that choice.

According to Einstein, flat space-time manifolds are those entirely free of gravitational fields; as a class they are not those which one should expect to serve as a foundation for a theory of gravitation. Some other class of manifolds must play that role. The principle of general covariance implies that this class is one that does not lead to a privileged or special class of coordinate systems. Rather, the geometric relations characteristic of these manifolds are to be such that they take the same form in all conceivable coordinate systems, and that they do not lend themselves to simplification in only some coordinate systems.

Reliance on the principle of general covariance drastically restricts the range of potential mathematical-geometric relations that might be considered as representing the laws of gravitation.

Also, the relativistic laws of gravitation should reduce approximately to Newton's law in the limiting situation in which the field is relatively weak (the masses of the sources of the field not too large) and the relative velocities of both sources and objects of the gravitational interaction are small compared to the speed of light c. These two requirements—that the theory satisfy the requirement of general covariance, and that it approach the laws of Newton under reasonable conditions—are sufficient to remove all ambiguities and to lead to a single possible law of gravitation. This is the theory that Einstein published in 1916.

11 Curved Space-Time

To describe and to explain Einstein's theory of gravitation, some further discussion of curvature of spaces is necessary. As indicated previously, a space is said to be *curved* if the result of parallel transport of a vector from one point to another depends on the path chosen to connect these two points with each other. Hereafter, it is accepted that parallel transport is to be considered a well-defined operation along any chosen curve, which leaves unchanged both the length (or magnitude) of the vector being transported and the angle between two vectors which are transported along the same path.

If the vector to be transported happens, at some point along the path, to be tangential to the path itself, then it depends on the choice of path whether parallel transport preserves this property all along the path or not. In the former case the chosen path is said to be an *auto-parallel* (self-parallel) curve. Given a point and an initial direction, there is always exactly one auto-parallel curve that passes through the given point in the given direction. On the surface of a sphere the great circles are the auto-parallel curves. In a flat space, they are the straight lines.

If the result of parallel transport from one point to another depends on the choice of connecting path, then parallel transport from a point along a closed path back to the point of departure usually will result in a vector that differs from the original vector. Because, by assumption, the magnitude of the vector is to remain unchanged in the course of parallel transport, transport along a closed path will at most result in a rotation, but no stretching. If a set of vectors is taken along a closed path together, the whole set is turned as a rigid configuration, because no angles between vectors are to change, either.

In spaces of several dimensions, curvature is complicated by considerations of orientation. To think of curvature as a local property of a space, involving primarily the immediate vicinity of any given point of the space, one considers small loops along which parallel transport is to be performed. If the loop is sufficiently small, the angle through which a vector is turned is proportional to the area bounded by the closed path and independent of the shape of the path. Thus the appropriate measure for curvature is the turning angle per unit area circumnavigated. This measure will, however, depend on the orientation of the surface in which the closed path is embedded, an orientation that may be described conveniently in terms of any two linear directions tangential to the surface. In a four-dimensional continuum, such as the space-time manifold, there are six independent possible orientations for a two-dimensional surface, in the sense that any possible orientation may be built up from six basic ones. In a Minkowski space-time, and with a standard Lorentz frame, these six basic orientations are the ones spanned by all combinations of the directions of the four coordinate axes: (xy), (yz), (zx), (xt), (yt), and (zt).

Aside from the orientation of the area enclosed by the loop, the turning angle also depends on the direction of the vector that is being transported. The amounts and directions of turning of different vectors that are taken about the same closed path are, however, not independent of each other but related by the requirement that any configuration of several vectors turns as a rigid whole. In a four-dimensional continuum there are, again, six basic independent ways of rigid rotation, from which all others may be built up. Thus it would seem that there must be $6 \times 6 = 36$ basic components of the curvature, corresponding to all possible orientations of the closed path of transport and all posssible ways in which a rigid assembly of vectors can be turned. The actual number of basic components of the curvature is, however, smaller, because of additional considerations, and amounts to twenty components that are truly independent of each other. These additional considerations involve a partial interchangeability of directions, too technical a matter for discussion in this book.

The twenty components of the curvature in a four-dimensional space can be grouped into two sets of ten components each, in a manner that is independent of any choice of coordinate system. One of these two sets involves the turning of vectors in the course of parallel transport in a surface that is spanned by the vector to be turned and one other, fixed vector; this set is usually referred to as the Ricci tensor, named for the Italian mathematician, Curbastro Gregorio Ricci (1853–1925). In a slight rearrangement of the same components, the Ricci tensor may be called the Einstein vector. The remainder, the other ten components, form an entity, Weyl's tensor, named for the German-born mathematician, Hermann Weyl (1888–1955). The totality of all components of the curvature is called *Riemann-Christoffel's* curvature tensor to commemorate two mathematicians—Georg F. B. Riemann (1826–1866), a German, and Elwin Bruno Christoffel (1829–1900), a Swiss—who pioneered in the study of the respective combinations.

Besides the separation of the curvature into components, one is concerned with the strength of curvature to be anticipated in a given physical situation. Instead of the angle per unit area, the normal measure of curvature, one may adopt a somewhat more intuitive measure, the size of a sphere whose surface has a given amount of curvature. The smaller the sphere, the greater its curvature. The unit of curvature is represented by a surface on which a vector of unit length turns one radian when transported around the boundary of a square of unit side length. The *radian* is that angle that, when subtended at the center of a circle, cuts out an arc segment of the circle whose length equals the circle's radius. One radian equals about 57°. The sphere whose surface has a curvature of magnitude 1 is the sphere whose radius equals 1. The sphere whose radius equals 2 has a surface with a curvature of $\frac{1}{4}$. The curvature of a sphere whose radius is R will amount to $1/R^2$.

Consider now the gravitational field on the surface of the earth. The magnitude of the gravitational acceleration is, in round numbers, 10 m/sec^2, and this is the acceleration that a free-falling frame should have relative to the earth. If such a frame is extended over some distance, other free-falling test bodies will be observed to have some acceleration with respect to that frame, and this acceleration will be proportional to the distance from the site where the construction of the free-falling frame was begun. As the radius of earth is some 6,000 km, it follows that the acceleration relative to the free-falling frame changes very roughly at the rate of 10 m/sec^2 per 6,000 km distance from the original site, or 1.6×10^{-6} sec^{-2}.

Adopt now as a closed path for parallel transport within the free-falling frame a rectangle of which one pair of opposite sides is spacelike and vertical, 1 m long and parallel to x, the other pair timelike and 1 sec long. Have one of the two timelike sides placed at the origin of the free-falling frame, the other timelike side 1 m away from the origin, where the apparent gravitational acceleration will be 1.6×10^{-6} m/sec^2 (Fig. 40). Consider now as the vector a to be transported about this rectangle a vector paral-

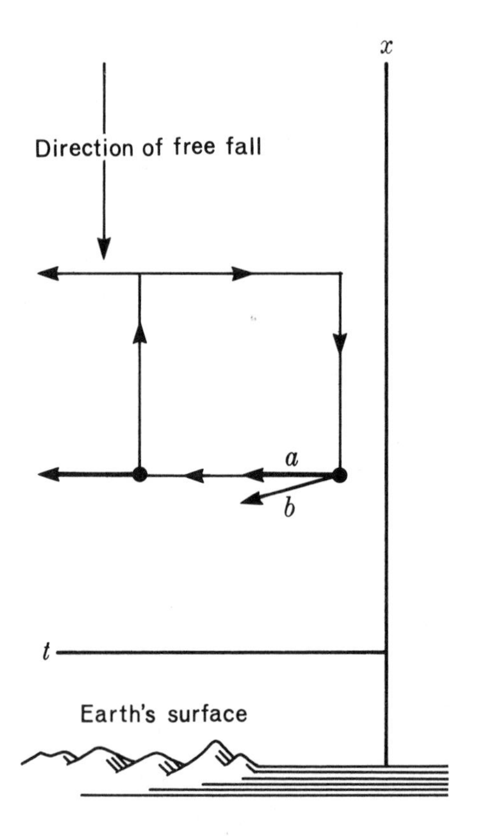

Fig. 40. Curvature of the Earth's Gravitational Field

lel to the time axis t, such as the velocity vector of a test body at rest. Transported first along the time axis, this vector remains unchanged (with respect to the free-falling frame), as a test body initially at rest remains at rest. Transport along a spacelike side has no effect either, but transport along the third, timelike side results in a change in velocity by an amount of 1.6×10^{-6} m/sec^2, leading to b. Transport along the fourth, spacelike side produces no further change.

Thus, parallel transport about a loop that encloses an area of 1 m sec produces an amount of turning of a vector of length 1 by an amount of 1.6×10^{-6} m/sec. The curvature is the ratio between the amount of turning and the area circumnavigated, or 1.6×10^{-6} sec^{-2} or, in units of m^{-2}, about 1.5×10^{-23} m^{-2}. Conversion from seconds to meters has been accomplished by using the value of the speed of light, 3×10^{8} m/sec. This curvature, then, is about the same as that of the surface of a sphere whose radius is roughly 1,300 light seconds, or twenty light minutes, a little more than the diameter of the earth's orbit about the sun.

Recently, with the advent of satellite techniques, an experiment has been proposed to make the curvature of space-time directly observable. Suppose one places a gyroscope into orbit; then its axis will point in the same direction in space with considerable stability. This property makes gyroscopes such useful navigational devices. Ships' compasses have been gyroscopes (rather than magnetized steel needles) on all large vessels for a good many decades. As for the gyroscope aboard an artificial satellite which circles the earth, its axis would be transported bodily about a closed path; assuming the soundness of Einstein's conjecture that space-time is curved as a result of gravitation, the axis of the gyroscope should be turned slightly after each such circuit, but only less than one one-hundred-millionth (10^{-8}) of a right angle. If a low-level satellite, which circles the earth in about 1.5 hours, continues its circuits for a year, the turning angle will add up to about 5 seconds of arc, or a bit better than one hundred-thousandth (10^{-5}) of a right angle, and such an angle is barely observable. The experiment will be difficult to instrument and to perform cleanly, because there are a number of extraneous influences that also tend to turn the axis of a gyroscope; these extraneous influences must be either eliminated or calculated precisely if they are not to mask the effect looked for. (Basic instrumentation for the experiment is currently being developed in two laboratories in the United States.)

12 Gravitation in the Space-Time Continuum

If all twenty components of Riemann-Christoffel's curvature tensor were to be required to vanish, then the space-time continuum would be flat, and there would be no possibility of a gravitational field. If, on the other hand, there were no restrictions on the curvature tensor, any sort of accelerations could be realized as gravitational fields, contrary to the observed fact that gravitational fields obey definite laws, approximately those discovered by Newton. The laws of gravitation must somehow restrict the curvature but without ruling out curvature altogether.

As mentioned previously, the twenty components of the Riemann-Christoffel tensor may be split into two tensors with ten components each. In the geometry of curved spaces, one considers transitions from one coordinate system to another and their effects on structures consisting of several components. Take an ordinary vector V. Because the vector is not merely a quantity but possesses the elements of both magnitude and direction, it is fully represented only by four components in a four-dimensional space (V_x, . . . V_t in Fig. 41). These components represent, as it were, the respective projections of the whole vector into the four directions represented by the coordinate axes. If one coordinate system, x and t, is replaced by another, x' and t', the new coordinate axes will have directions different from the former axes; hence in the new coordinate system the same vector will be represented by a different set of four components, $V_{x^1}, \ldots V_{t^1}$. Only the dimensions x and t are shown here. But if the relationship between the original and the new coordinate system is known (the so-called coordinate transformation), then there is a definite rule for determining arithmetically the values of the new vector components from the values of the old ones. This rule is known as the *transformation law* for vector components. Similar transformation laws apply to other geometric structures, such as tensors.

Given a fairly involved structure, such as the curvature tensor, the transformation law for its components under coordinate transformations is fairly involved, too, but it is quite unambiguous. If all the twenty components of the curvature tensor are known in one coordinate system, they may be computed in any other coordinate system. But not all the twenty components in one coordinate system are needed to find the values of any one of the

Fig. 41. Components of a Vector in Two Lorentz Frames

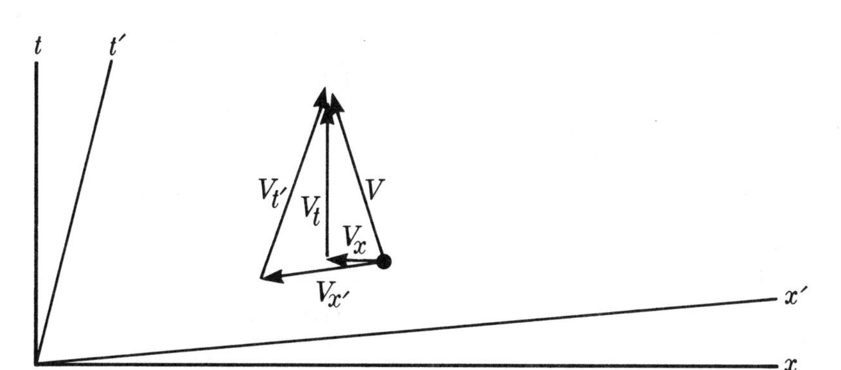

new components. Rather, there are smaller sets of components knowledge of which in one coordinate system suffices to obtain the corresponding components in another system. These are the two sets of ten components each, which are known as the Ricci (or Einstein) tensor and as Weyl's tensor, respectively. If the ten components of the Ricci tensor vanish in one coordinate system, they will be zero in every other coordinate system, and the same holds of Weyl's tensor.

Weyl discovered an important property of the tensor named after him, its *conform-invariance.* In all preceding discussions it was assumed that the geometry of a given space-time is characterized by the invariant interval between any two world points, and that this interval is not to change its numerical value under coordinate transformations. Indeed, coordinate transformations involve no other changes than relabeling world points; they are not to affect the underlying geometric properties of space-time. Aside from coordinate transformations there are some changes that affect the geometry itself; that is, operations that change one space-time manifold into another manifold. Weyl considered in particular *conformal transformations.* In a conformal transformation all intervals between close pairs of points are changed by multiplication by an arbitrary factor, which may be different in different locations but which, at one location, is independent of the direction in which the second (nearby) point is located in reference to the first point. In a conformal transformation, angles between curves at the point of intersection remain unchanged. In making maps, the so-called stereographic projection (Fig. 42) is a well-known conformal transformation, which changes the surface of a sphere (the globe) to a plane (the map).

If a conformal transformation is performed on a four-dimensional space-time continuum, its geometry will be changed; for instance, a Minkowski (flat) continuum may be converted to a

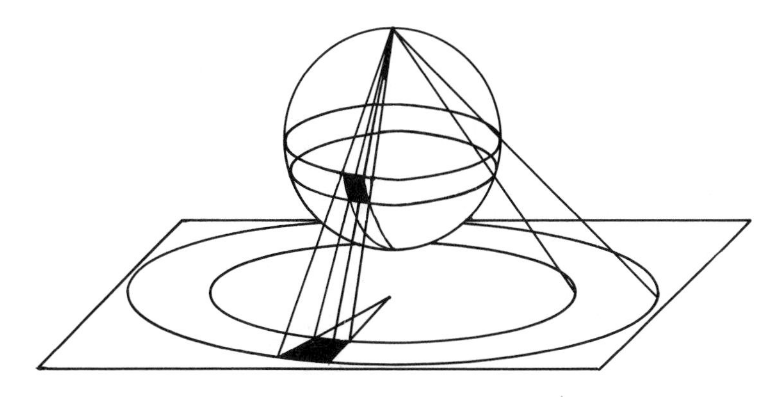

Fig. 42. Conformal Mapping (Stereographic Projection)

curved continuum. But one aspect of the space-time geometry remains unchanged by conformal transformations, and that is the spacelike, timelike, or lightlike character of each interval and direction. Particularly, the two light cones of directions at each world point remain unchanged by any conformal transformation, and that is why Weyl was interested in studying them. He discovered that, under a conformal transformation, the Ricci tensor and the Weyl tensor behave very differently: the Ricci tensor changes in a very complicated way (which includes the possibility that it may vanish before the conformal transformation but be non-zero after the transformation); the Weyl tensor remains unchanged. Hence it is sometimes also called the *conformal curvature tensor*.

In Newton's theory of gravitation, the gravitational acceleration caused by a given large mass is proportional to that mass, and inversely proportional to the square of the distance from that mass. The same law may also be formulated in a slightly different manner, which is capable of providing a clue for the form of the relativistic law of gravitation. This alternative formulation is based on the notion of the *gravitational field* as something imprinted on

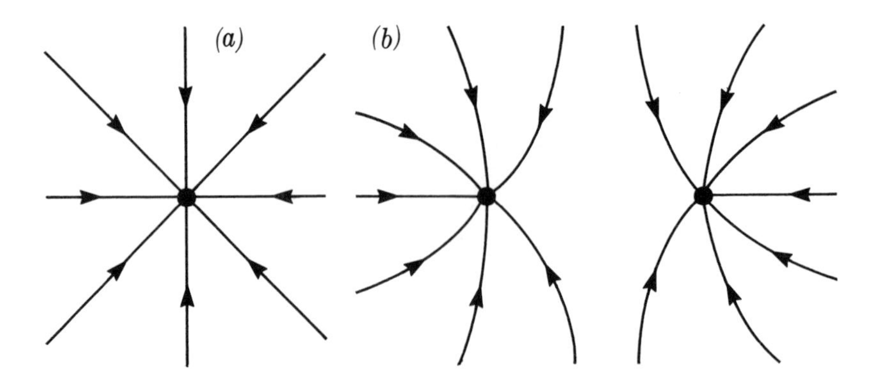

Fig. 43. Lines of Force

the surroundings of the gravitating large mass, irrespective of the presence or absence of any test bodies. The field is fully described by the vector whose magnitude and direction at each point in space represent the gravitational acceleration that would be experienced by any test body taken to that location. One can describe the gravitational field graphically by means of curves which, at each point in space, have the direction of the local gravitational field (acceleration) and which are distributed in a density (such-and-such a number of curves per unit of cross-section area; Fig. 43) equal to the magnitude of the local field. If only one large mass is present, these curves will be straight lines and directed straight at the gravitating mass, as shown in Fig. 43a. Fig. 43b corresponds to two masses.

The inverse-square law now amounts to the principle that all of these lines originate at infinity (at indefinitely large distances from the region of interest) and terminate in the large mass. For if the density is to be equal to the magnitude of acceleration, then the number of lines passing through a spherical surface centered on the large mass will be equal to that density, multiplied by the area of the spherical surface of radius r, and that latter is proportional to the square of the radius. Hence the product of density and area, the number of traversing curves, or lines, is the same for each of the concentric spheres. And the total number of lines through each spherical surface surrounding the large mass must be the same; this constant number is proportional to the mass that is the source of the field. Quite generally, Newton's inverse-square law may be put into a form that accommodates either one large mass as the source or any distribution of masses: All curves of the gravitational field originate at infinity and terminate at masses. The number of curves terminating in any region filled with masses is proportional to the total mass contained in that region. Secondly, the gravitational field is *conservative:* curves cannot form loops, and no energy can be gained or lost by moving a test body along any closed path.

In the relativistic theory of gravitation, the role of the sources is necessarily assumed by the combination of mass (or energy) and linear momentum. The distribution and flux of energy and linear momentum is represented by the energy-stress tensor, with ten components, which satisfies four laws of continuity (Chapter 6). In any one frame of reference, one of these four laws of continuity is associated with the conservation of (relativistic) mass, the other three, with conservation of the three components of linear momentum. But just as the mass in one frame of reference is determined not by the mass alone but only by the mass plus linear momentum in another frame, so the four laws of continuity together form a vector law, and the validity of any one of them in one frame of reference is assured only by the validity of all four of them in another frame.

The Newtonian law on curves of the gravitational field is a law concerning the inhomogeneities of the field; the requirement that no curves originate in free space places definite limitations on the way the field changes from point to point, not on the magnitude of the field itself. Moreover, the phrasing of the law in terms of properties of the curves of the field has the virtue of dealing with local properties, not with the locations of distant sources. The distant sources determine the local field, to be sure; but they do so by the intermediary of determining how many lines of force must originate at infinity in order that the right numbers may terminate in the sources.

Inhomogeneities of the relativistic gravitational field are described by the curvature tensor, whereas the field itself is described by the manner in which a vector, such as the four-velocity of a particle, is to be changed along a curve (its trajectory in space-time). For this reason it appeared likely to Einstein that the relativistic law of gravitation would relate the curvature tensor, or some of its components, to the tensor representing the behavior of the sources. Which of the components of the curvature should be related to the sources was at first obscure, until Einstein discovered that the tensor now called after him, a slight rearrangement of the Ricci tensor, satisfies four laws of continuity, with a formal structure exactly like that of the energy-stress tensor. Accordingly, he postulated that these two tensors are proportional to each other. The constant of proportionality was determined by the requirement that the new law should reproduce Newtonian physics in the limit of weak fields and small velocities; it is essentially equal to Newton's constant of gravitation.

With this step, Einstein completed the foundations of his theory of gravitation, also known as the *general theory of relativity.* To sum up, the theory involves a generalization of Minkowski's geometric notion of a four-dimensional space-time manifold, in that gravitational fields are interpreted as manifestations of the curvature of that manifold. Specifically, the acceleration of test particles relative to an arbitrary (not free-falling) frame of reference corresponds geometrically to the deviation of auto-parallel curves (which are identical with the geodesic curves) from the chosen coordinate axes. Such deviations can occur whether the manifold is flat or curved, depending on the choice of coordinate system, and this geometric fact is again interpreted physically in terms of the similarities between inertial and gravitational accelerations. True gravitational accelerations are distinguished from inertial accelerations because no choice of frame of reference can eliminate truly gravitational accelerations everywhere; in geometric language, if the manifold is curved, then there is no coordinate system whose axes are auto-parallel (or geodesic) everywhere.

Given the identity of the physical notion of the true gravitational field with the geometric concept of the curved manifold, the laws of gravitation were bound to be represented by limitations on the curvature of the manifold. As gravitational fields are caused by the presence of large masses, and as masses in a relativistic theory are best represented by the ten-component energy-stress tensor, the most natural form of any law of gravitation is some relationship between the energy-stress tensor and the curvature. The Einstein tensor commanded attention, because its transformation laws are the same as those of the energy-stress tensor (not all tensors have identical transformation laws, but these two do) and because, just as the energy-stress tensor, it satisfies four laws of continuity. Hence, if one relates the gravitational field to the known sources of gravitation, the masses, by requiring proportionality between the energy-stress tensor and the Einstein tensor, then this proportionality will hold in all conceivable coordinate systems, and in exactly the same form. Thus the principle of general covariance is satisfied. Finally, Einstein was able to show that in situations in which the massive bodies move relative to each other at non-relativistic velocities (this means at relative speeds representing but a small fraction of the speed of light), and provided that the curvatures are slight (this means that the radii of curvature are large everywhere compared to the distances between the interacting masses), his laws of gravitation lead to results nearly indistinguishable from those of Newton's theory of gravitation.

In the typical astronomical situation in which the large masses interact with each other across vast empty spaces, the energy-stress tensor is non-zero only inside the various constituent massive bodies but vanishes elsewhere. The Einstein tensor, too, must be zero everywhere outside the stellar interiors. But as the total curvature at any given point is composed of both Einstein's and Weyl's tensors, it will be non-zero even where Einstein's tensor vanishes. Thus there is a true gravitational field even in the empty spaces separating the gravitating masses, as required by the known facts.

Einstein had constructed his law of gravitation so as to incorporate the laws of continuity of energy and of linear momentum. For any isolated physical system, conservation of linear momentum is tantamount to preservation of the velocity of its center of mass. Such a system obeys Newton's law of inertia. Thus, though the general theory of relativity denies the validity of the notion of inertial frames of reference in the presence of gravitational fields, isolated systems when viewed from afar obey the principle of inertia, which originally had given rise to the notion of inertial frames. If the universe were to consist of a condensation of matter surrounded by infinitely extended regions of empty space and of diminishing gravitational fields, then general relativity would permit the construction of an inertial frame at least asymptotically in that far-off region. Present astronomical evidence, however, does not favor the idea of such an "insular" universe.

Because of the extreme slightness of the curvature of space-time, the predictions of the general theory of relativity match very closely those of a theory patterned after the (special-) relativistic theory of electromagnetism. The principle of equivalence, according to which gravitational and inertial masses equal each other for all physical objects, is, however, built into its foundations. This principle cannot be eliminated without destroying the theory as a whole. For slow motions and for moderate gravitational fields, the new equations resemble the formulas of Newton's earlier theory in such close approximation that so far only one minute difference between the two theories has been observed astronomically, the advance of the perihelion of Mercury.

A number of other effects predicted by the new theory should be observed under extreme conditions, when large bodies move at velocities approaching that of light, or in the presence of masses so large that the curvature caused by them has a radius not much larger than the dimensions of the massive body itself. None of these effects have as yet been observed, but astronomers have advanced speculations that such conditions may be met with in other parts of the universe.

13 Schwarzschild's Solution

Within months of Einstein's publication of the proposed new laws of gravitation, the German astronomer Karl Schwarzschild (1873–1916) in 1916 reported the first "rigorous" solution of these equations, a solution that was not approximate and that had been obtained independently of any assumption that the fields were weak. Schwarzschild's solution represented the gravitational field of a single spherical mass in the surrounding space: at sufficiently large distances, the solution assumed the characteristics of the classical inverse-square law of gravitation. In fact, if the source of the gravitational field were a celestial body of moderate total size and moderate density, then Schwarzschild's solution would be all but indistinguishable from Newton's. Only if the mass of the source were so highly concentrated as to permit "large" gravitational fields on its surface would one get interesting new effects, which might be observed.

But what is a "large" field? If anything is to be learned from astronomy and physics, it is that man and his experience are a poor yardstick by which to judge what is "large," what "small." Surely, a celestial body is large compared to our everyday notions, but the earth is small compared to a fixed star, and a star small compared to a galaxy. Again, man is unbelievably large compared to a bacterium, a bacterium large compared to a virus, and a virus much larger than an atom. Thus the gravitational field met with on the surface of the earth is not to be accepted as a universal standard of gravitational field strength. In any case, what is substantive about the gravitational field is not so much the field strength itself—the acceleration of a test particle at a given location (for that depends on the choice of four-dimensional coordinate system)—but the resulting curvature (Fig. 44). The curvature in turn can be described in terms of the radius of curvature (the radius of a sphere possessing the same curvature), being large if the radius of curvature is small, and vice versa. Gravitation is to be considered "strong" if the radius of curvature is not very large compared to the geometric dimensions of the object considered. If all the mass of the earth were concentrated at one point, so that gravitation became steadily stronger as one approached that center, the radius of curvature of space-time would approach the distance from the center of the concentrated mass at 1 cm (four-tenths of 1 inch). If, similarly, the mass of the sun could be concentrated at one point, the curvature would become considerable, in this same sense, at a distance of about 1 km (three-fifths of a mile) from the center. In both these situations, one refers to the region in which the curvature becomes that large as the *Schwarzschild radius* (Appendix IV) associated with the mass of the earth or of the sun, respectively.

One can envisage the Schwarzschild radius in a slightly different fashion. Ever since men began to lift rockets off the surface of the earth, the notion of escape velocity has become popular. *Escape velocity is* the speed that must be imparted to a space vehicle to enable it to leave the vicinity of the earth in free (unpowered) flight (Fig. 45). The Schwarzschild radius is that region in which the escape velocity approaches the speed of light. This radius

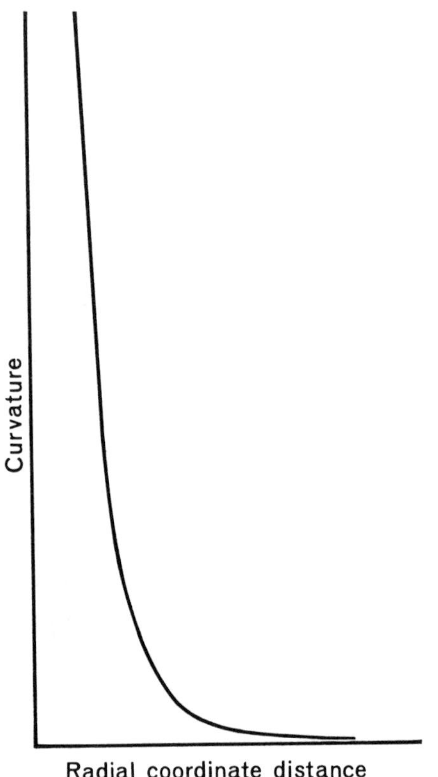

Fig. 44. Curvature of Schwarzschild's Field

can be calculated without reference to relativity at all. (In order not to burden this account with mathematical detail, this calculation has been relegated to Appendix IV.) The Schwarzschild radius R of a mass of magnitude M is given by the expression

$$R = \frac{2\kappa}{c^2} M .$$

As usual, c stands for the speed of light; κ is known as *Newton's gravitational constant,* the force with which two masses of 1 gm each will attract each other if separated by a distance of 1 cm.

This force is less than one ten-millionth of a *dyne*, the force that causes a mass of 1 gm to gain a velocity of 1 cm/sec in 1 sec. What would happen if one could approach the Schwarzschild radius of a sufficiently concentrated mass? This question has aroused considerable interest. Figures 46 and 47 indicate that, with a sufficiently large total mass, the average mass density need not be excessive in order to bring the Schwarzschild radius out to the surface of the star and into empty space. In Fig. 46, the Schwarzschild radius is plotted against the radius of a mass concentration whose mean density equals that of water under ordinary conditions. The two radii become equal to each other when the total mass equals 10^8 solar masses. In Fig. 47, the density required to bring the Schwarzschild radius out to the surface is plotted against the total mass of the star; for the mass of the sun, this density would amount to 10^{16} times the density of water, and

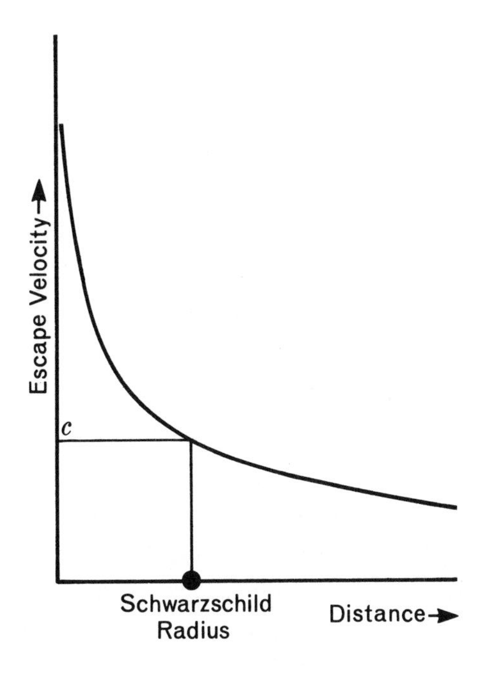

Fig. 45. The Escape Velocity as a Function of Distance from a Large Point Mass (Newtonian Mechanics)

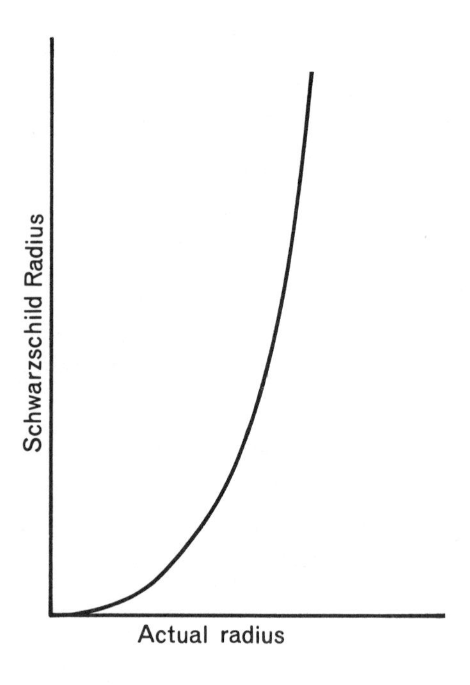

Fig. 46. Schwarzschild Radius versus Actual Radius of Density Taken as Constant and Equal to that of Water

that is approximately the density of matter inside an atomic nucleus. To bring the required mass density down to that of ordinary matter, the total mass required is of the order of a galaxy. In a real galaxy, however, the matter is distributed over a region so large that nowhere does the curvature of space-time reach a large value. Conceivably, the *quasi-stellar objects* which were discovered in 1963 by C. Hazard, an Australian, and a team of California astronomers, and some of the *interlopers,* which Allan Sandage and his group at the Palomar Observatory discovered a year later, might represent concentrated masses so large that the resulting curvature would become appreciable. *Quasars,* or quasi-stellar objects, are very bright objects, with large red shifts of their spectra, which emit strong radio waves. *Interlopers* resemble quasars in the range of visible light, but lack their intense radio-frequency emission. Currently, the distances, and hence the sizes,

Fig. 47. Density versus Mass if the Schwarzschild Radius is to Equal the Actual Radius

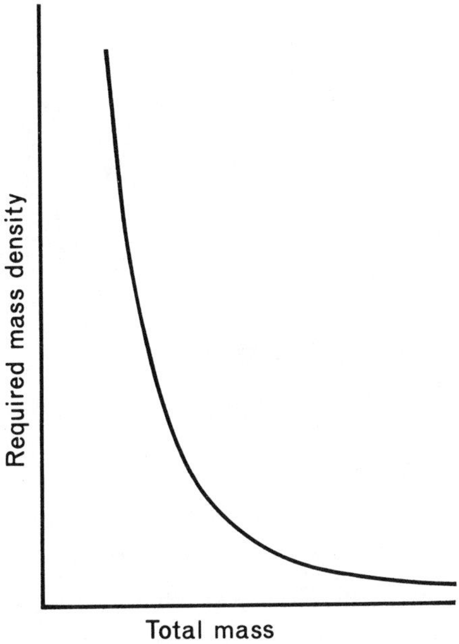

of quasars are in dispute; only when their masses and sizes are reliably known can the space-time curvatures on their surfaces be estimated with confidence.

Aside from the question of realization in nature, consider the hypothetical situation in which a large mass is sufficiently concentrated so that its Schwarzschild radius appears in the open, or is at least close to the surface. What kind of physical phenomena does the general theory of relativity predict? As long as observations are confined to regions far away from the massive body and from its Schwarzschild radius, the mass will act pretty much as it does according to Newtonian physics: it will attract smaller bodies, which will travel either on open trajectories coming from infinity and returning to infinity, or on elliptical trajectories. According to relativity, however, these ellipses do not themselves remain stationary, as they would in Newtonian theory, but will slowly turn in

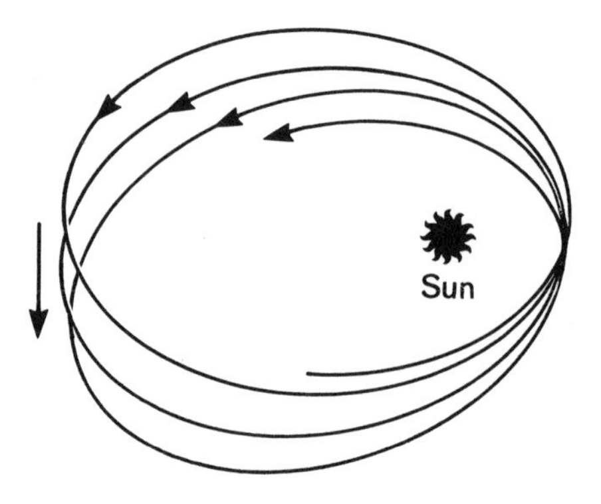

Fig. 48. Elliptic Orbit with Perihelion Advance

space in the course of a great many revolutions (Fig. 48). This slow turning, the so-called *advance of the perihelion,* has been verified for the orbit of Mercury, the only one of the planetary orbits for which the theory predicts a rate of advance sufficiently large to be observed. Very recent work by Dicke appears to indicate, though, that agreement between theory and observations of Mercury's perihelion advance may not be as good as it was thought to be originally.

Further, the theory predicts that the spectral lines emitted by atoms near a large mass will exhibit somewhat lower frequencies than the same lines emitted elsewhere. This is a special case of the more general assertion that all processes occurring near a large mass appear slowed down. The change in spectral frequencies is proportional to the Newtonian potential, and hence inversely proportional to the distance from the center of the large mass; it is known as *gravitational red shift,* because a decrease in the frequency of light shifts its color toward the red end of the spectrum. Spectral red shifts and blue shifts may also be brought about by diverse other causes, such as rapid motion of the light source (known as *Doppler shift*). The adjective "gravitational" identifies red shift caused by the location of the source within the gravitational effects of a massive source. In 1960, R. V. Pound at Harvard University succeeded in demonstrating in the laboratory gravitational red shift caused by the earth's gravitational field. Previously, this effect had been observed in the spectra of the very dense stars known as white dwarfs. It had also been looked for in the sun's spectrum, where the effect is so small that it is partly overshadowed by many other peculiarities of the light emitted by the sun's radiant atmosphere.

If it were possible to bring a source of light close to a concentrated mass, the slowing of its frequency would develop into a complete standstill at the Schwarzschild radius. This effect cannot, of course, be demonstrated experimentally in the absence of masses of sufficient concentration, but there is no reason to doubt that such a standstill would, in fact, be observed.

Distances would also be affected in close proximity to the gravitating mass. Suppose one were to place concentric spherical shells about the gravitating mass and identify each shell by the area of its surface. In ordinary flat space, the area of a spherical surface equals the square of the sphere's radius, multiplied by 4π. Conversely, knowing the respective areas of two such shells one can calculate the distance between them. This relationship is disturbed close to the Schwarzschild radius in that the distance between neighboring spheres is greater than it would be in the absence of curvature. Because it is difficult to illustrate a three- or four-dimensional curved space, Figs. 49 and 50 show a similar relationship for a curved two-dimensional space, a sphere on which concentric circles are shown. Distances between circles of increasing circumference are greater than in the plane. (The circumference of a circle plays the same role here as the surface area of a sphere in three dimensions.)

Consider now the free fall of a particle from some distance far from the large mass inward, and all the way to the Schwarz-

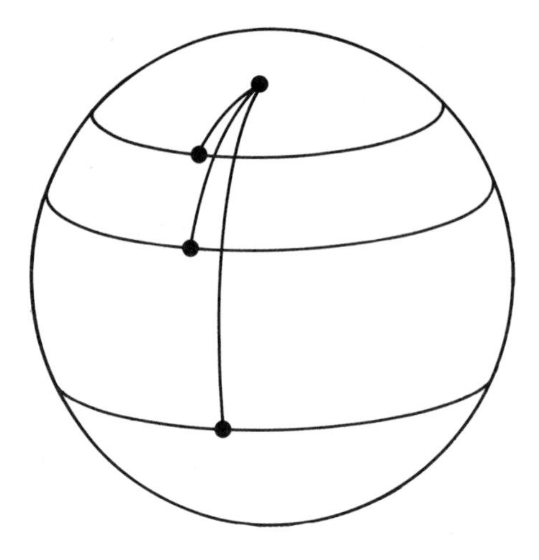

Fig. 49. Circles and Their Radii on a Sphere

schild radius. Starting from rest, this test body initially will build up its speed as it falls toward the large mass. It is reasonable to measure its speed in terms of the concentric spherical shells that the test body passes on its journey and to employ units of time as measured by an observer who remains outside and at rest. After a while, the speed of the falling particle will no longer increase;

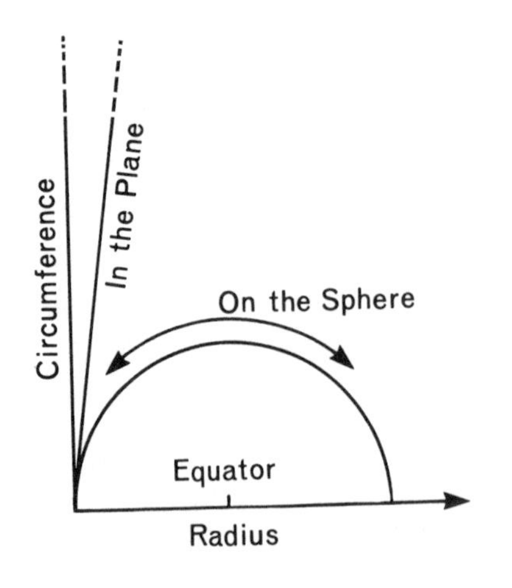

Fig. 50. Circumference of a Circle as a Function of its Radius in the Plane and on the Sphere

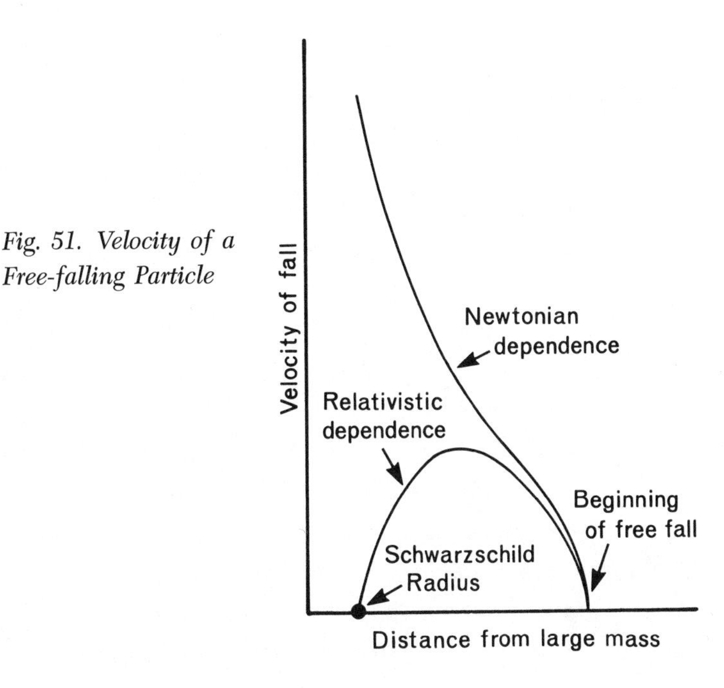

Fig. 51. Velocity of a Free-falling Particle

on the contrary, it will slow down because the motion of the particle is governed by local conditions. As it approaches the Schwarzschild radius (Fig. 51), the particle must traverse increasingly larger distances to pass from one concentric shell to the next; in so doing, it is controlled by local time, which passes more slowly than the outside observer's time. These two effects reinforce each other; they result in such a slowdown that it takes the particle *forever* to reach the Schwarzschild radius.

Not only particles but also light rays experience a slowdown. The theory predicts that a pulse of light originating from anywhere outside the Schwarzschild radius of a large mass will take an infinite time to reach it. If an experimenter could station a mirror somewhere near the Schwarzschild radius and then shine a light pulse from his outside location on that mirror, he would have to wait a long time before seeing the reflection of that flash, and the time lapse would increase without limit if he were, in subsequent trials, to place the mirror closer and closer to the Schwarzschild radius. By the same token, an outside observer would never be able to see an event that took place inside the Schwarzschild radius, because the light signal emanating from such an event would not reach him within any finite time.

The motion of material bodies and of light signals looks this way to an outside observer who himself remains stationary. But what about an observer attached to a freely falling particle? Strangely enough, such an observer takes only a finite time to reach, and to cross, the Schwarzschild radius (Fig. 52). At first sight, this result may appear contradictory. It is to be explained by the somewhat ambiguous use of the notion of "time." The observer who himself participates in the free fall will use as his standard of time a clock that he carries with him into the region

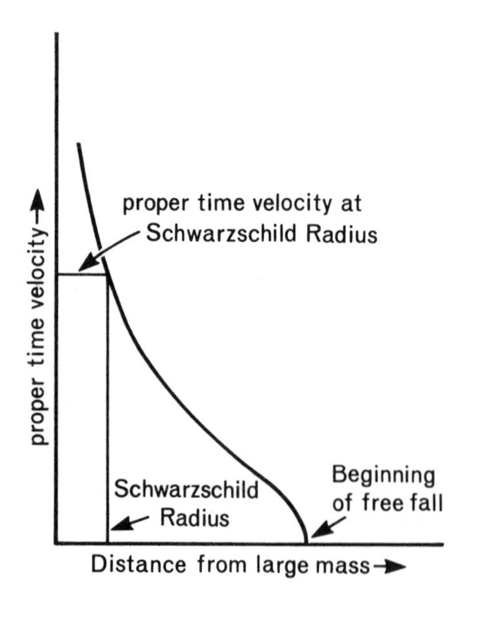

Fig. 52. Free-fall Velocity in Terms of Proper Time

in which all processes are slowed down. He will find that he reaches the Schwarzschild radius at a time in which the hands of his clock have traversed only a finite part of the dial. On the other hand, the stationary observer who wishes to judge the rate of a fixed clock which is closer to the Schwarzschild radius than himself simply looks at that clock and compares it to his own. He observes that his own clock's hands have traversed the dial several times by the time that the distant hands have completed one circuit. Thus he finds the distant clock (which is not falling but retains its location) to be slow.

To find the time of transit of a falling particle across a given spherical shell (or other marked point) the fixed observer can employ one of several stratagems. He may make a preliminary set of observations with the help of mirrors that will tell him how long it takes light to travel back and forth between him and any particular point farther inside. The one-way travel time is then taken to be half the time of the round trip. Next he proceeds to observe the free fall of a particle visually. Finally he corrects observed times of transit of the particle for the previously determined time it takes light to reach him.

Alternatively, the fixed observer may use as his particle a rubber ball that is perfectly elastic, so that it will bounce back from any fixed surface with undiminished speed. If he then places reflecting walls at various locations and determines the time that it takes his ball to fall toward a wall and to bounce back to him, he will have avoided employing light signals that must travel over large distances. Whichever method is employed, the results agree with each other: For a fixed observer, it takes a particle an infinitely long time to reach the Schwarzschild radius.

14 Inside the Schwarzschild Radius

Although, as measured by any outside observer, it takes a particle forever to reach the Schwarzschild radius, and although no light signals can cross the Schwarzschild radius in a finite time, free-falling objects take only a finite amount of their own time to reach the Schwarzschild region. Accordingly, it is not without interest to ascertain what adventures the theory predicts for the intrepid explorer who launches himself (rather than merely other objects) into the abyss.

Remember that this mythical project is meaningful within the framework of the theory only if a gravitating mass can be concentrated so highly that its Schwarzschild radius is found to lie in empty space. With any ordinary celestial body, none of the exotic effects associated with the Schwarzschild radius will be present. To illustrate this point, Fig. 53 shows the amount of red shift to be expected both when all the mass is concentrated at one point at the center (*a*) and when it is distributed through a finite volume extending beyond the Schwarzschild radius (*b*). It is not that extended matter interferes mechanically with investigation of the

Schwarzschild region, because instruments might not penetrate stellar matter. Even if a tunnel could be dug through the entire interior of an extended celestial body, one should not find any strange phenomena at the (nominal) Schwarzschild radius because not all the spread-out matter is effective in determining the curvature inside. Figure 54 illustrates the increase in the curvature of space-time as one approaches the center of an extended accumulation of gravitating matter, whereas Fig. 44 was drawn on the assumption that the mass was concentrated at the center.

Figure 55 shows conditions that will obtain if all the matter is contained at the point at the center, obviously a highly idealized situation under any circumstances. This illustration represents only one spatial dimension (a radial direction passing through the center) and the time. In this illustration, a variable scale has been chosen so that at each point light rays traveling either in or out are represented as straight lines at a 45° slope. Any direction that lies between one of these 45° slopes and the vertical will be timelike; any direction less steep than 45° will be spacelike. Points that have the same distance from the center do not lie on vertical lines but on curves that are shaped somewhat like

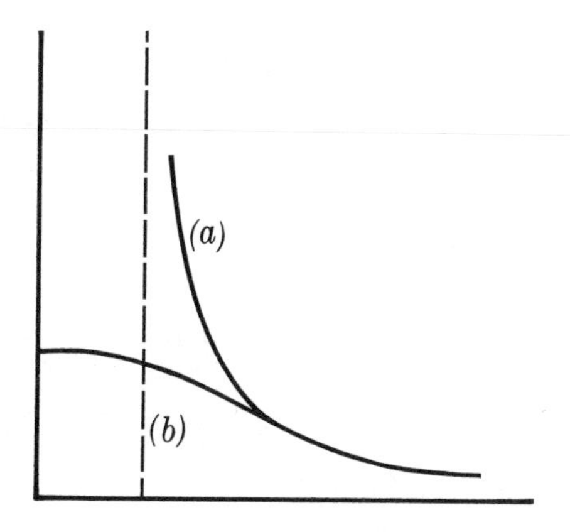

Fig. 53. Red-Shift Dependence on Distance

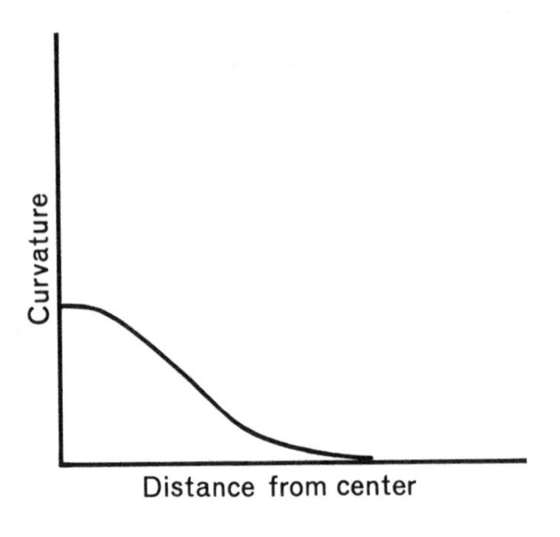

Fig. 54. Space-Time Curvature for an Extended Mass

hyperbolas. And points that represent "the same time" lie on curves all passing through the same point; this single point represents the Schwarzschild radius at all finite times. The two 45° lines jutting out from this point to the right represent the Schwarzschild radius at the infinite future and at the infinite past, respectively. They delimit the segment of space-time that might be

Fig. 55. Geometry near the Schwarzschild Radius

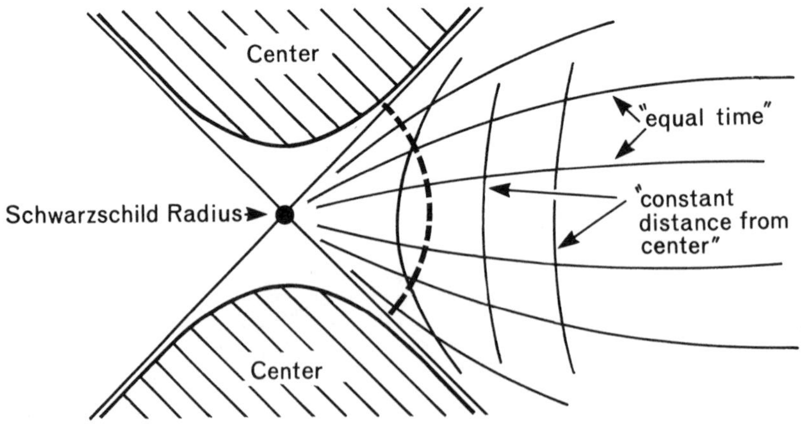

designated as the exterior of the Schwarzschild radius; this exterior region alone is accessible to two-way signaling from the outside.

The dashed curve represents a possible curve for a material particle, a trajectory whose slope is everywhere timelike. Because of the graphical limitations of Fig. 55 this trajectory represents purely radial motion, straight into the center or straight out. Part of this trajectory lies in the two-way accessible region, outside the Schwarzschild radius. Any one of the curves in the accessible region that corresponds to a constant value of r is suitable for the location of an observer stationary in time. Such an observer can send out light signals to any part of the trajectory of the material particle in the exterior region and receive its reflection back at a later time; he can maintain two-way communication with the material particle. Two-way communication breaks down at the two points at which the material particle crosses the Schwarzschild radius: one on its way to the exterior; the other, on its way back into the interior. The observer can view the instant of emergence from the interior, though he could not have sent a signal of his own to greet that event. Conversely, a signal emanating from the observer will reach the particle at the instant that it disappears behind the Schwarzschild radius, but there is no way that a signal marking the instant of disappearance can reach him.

The interior of the Schwarzschild radius contains two distinct parts, labeled Interior (past) and Interior (future). The stationary observer can see events taking place in the first (past interior) region, and he can dispatch messages toward the second (future interior), but not vice versa. No signal from the future interior can ever reach the exterior. A third part of the interior is completely inaccessible by light signals in either direction.

Because of the extreme distortion resulting from the variable scale of Fig. 55, the boundaries of the cross-hatched regions, which are marked "center," represent a single point, the center. It is not meaningful to talk of this particular point "in the course of time," as time does not have the usual significance in the interior. As far as the outside stationary observer is concerned, it takes forever just to reach the Schwarzschild radius; he cannot attach a meaningful notion of time to anything that goes on in the interior. He is insulated from the interior by what is known as an *event horizon:* He is unable to engage in two-way communication with any inside observer, by light signals or by other means. The outside observer either cannot see or cannot be seen, depending on whether his colleague is situated in the future interior or in the past interior. (Chapter 15 discusses event horizons further.)

What about the observer who plunges into and through the Schwarzschild radius? As has been pointed out, he will reach the crossing point within a finite amount of time; his wrist watch will show only a finite number of seconds elapsed from the starting point of his journey. Once he has crossed over into the future interior region, he will still be able to see the exterior region, though unable to send out signals of his own. He will not see any special effects as he crosses the Schwarzschild radius. His environment will look quite normal to him. As the outside observer approaches the center itself, the curvature of the space-time about him will continue to increase; it will become infinite as he reaches the center, and the center region will look "abnormal" to him in many respects. That he is unable to signal to the exterior once he himself is inside will not be apparent to him. Signals that he sends out will appear to spread normally. They fail to reach the exterior because the exterior appears to recede from him at the speed of light; hence his own light signals fail to catch up with the boundary. But as the boundary is not marked by any special properties, he does not observe the movement of the boundary away from himself. If he fixes his attention on his colleague, the stationary observer who has refused to accompany him on the way inside, he will notice that the observer's clock appears to go more and more slowly—but it does not stop. Conversely, if the stationary observer looks at the watch of his foolhardy friend, that clock will slow down and appear never to reach the point at which the falling observer crosses the Schwarzschild radius.

15 Event Horizons

Event horizons were unknown in prerelativistic physics, but they are normal in relativistic physics. They occur even in the absence of gravitational fields if two observers are accelerated in certain ways. Figure 56 illustrates the trajectories of two observers who are accelerated away from each other, with the coordinates x and t. Neither will ever be able to see the other if both are accelerated uniformly away from each other in such a manner that no lightlike line intersects both trajectories. If at any time the distance between the two observers is D, and each undergoes a constant acceleration, then the situation of Fig. 56 will be realized if the product of the acceleration and D is greater than twice the square of the speed of light. This condition will be satisfied if, for instance, the acceleration of each observer is about the same as the gravitational acceleration on the surface of the earth (32 ft/sec^2 or 10 m/sec^2) and if the distance of the two observers from each other exceeds two light years.

There are other kinds of event horizons. Again, in the absence of a gravitational field, if one of two observers is unaccelerated, whereas the other undergoes an acceleration, as illustrated in Fig. 57, the unaccelerated observer O_2 cannot see the accelerated observer O_1 at first but will be able to do so once he has passed the instant B. (The curve is a hyperbola.) He can make his presence known to his colleague, but only if he flashes a signal before instant A. Later on, he is invisible to his colleague. Between times A and B, he can neither receive nor send signals. On the other hand, the accelerated observer sees his colleague at all times, and he can send him signals as well; his observations are, however, confined to the unaccelerated observer's history before A, and all of his signals will be received after B.

To the observers involved, none of these event horizons appear as a sudden switching-on or switching-off of communication. For instance, when the unaccelerated observer begins, at B, to see the light held by the other observer, its apparent intensity is zero at first, though it rapidly increases to finite values. But the onset is always gradual.

Generally speaking, event horizons appear if, along the trajectory of some observer or of some object, either the past light cones or the future light cones fail to span all parts of the uni-

Fig. 56. Two Accelerated Observers

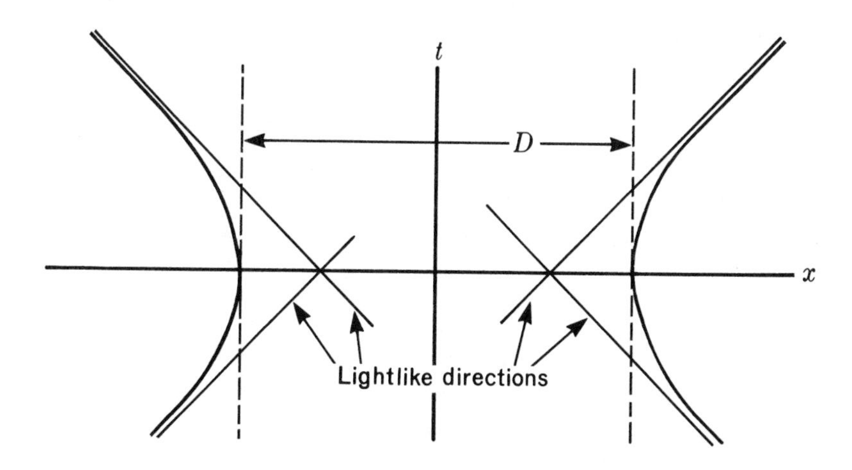

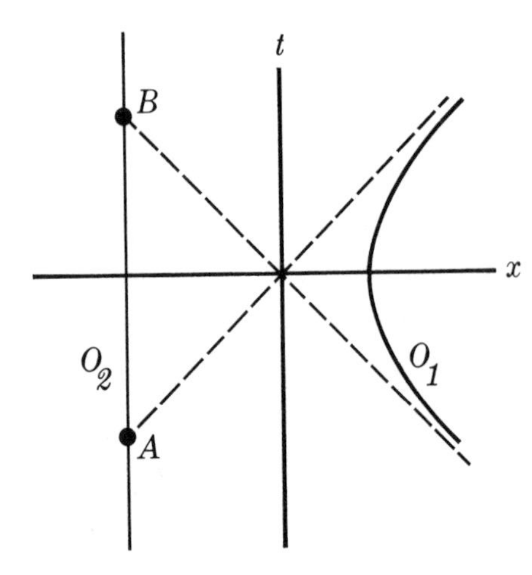

Fig. 57. Two Observers, One Accelerated, the Other Inertial

verse. In a flat Minkowski universe, the light cones associated with any inertial observer will in fact cover the Minkowski universe. Because the past light cones associated with an inertial observer have this power of coverage, such an observer will be able, sooner or later, to see every event no matter where or when it takes place. Because the future light cones cover the whole universe, an inertial observer, or an unaccelerated object, can be seen from anywhere at any time (Figs. 58 and 59). Figures 56 and 57 show that when an observer is non-inertial (accelerated), the associated light cones may fail to cover all of the universe, so that such an observer cannot see everything; similarly, an accelerated object may be invisible to some observers, some or all of the time.

In the presence of gravitational fields, the situation resembles that in Minkowski space-time but is complicated by the circumstance that an observer may be "at rest" but not free-falling. Under some circumstances, it may even happen that two free-falling observers are outside each other's horizons. In the presence of a Schwarzschild field, one caused by a single large mass, an ob-

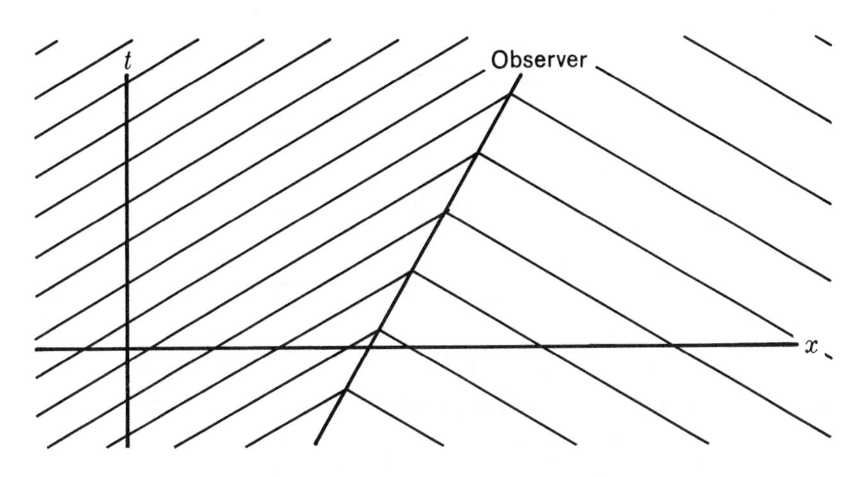

Fig. 58. Past Light Cones of an Inertial Observer

server at rest and outside the Schwarzschild radius cannot communicate with any part of the interior, whereas a free-falling observer can, if his motion is purely radial. There is, however,

Fig. 59. Future Light Cones of an Inertial Observer

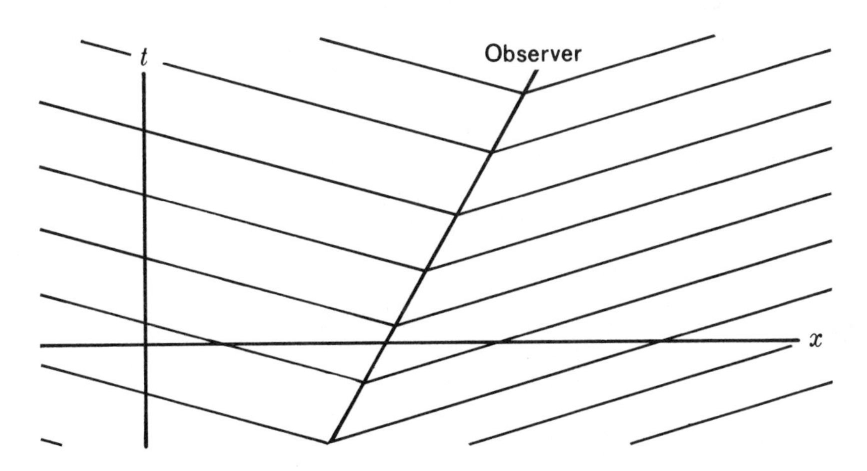

the not uncommon case of the observer on a circular orbit about the large mass who is free-falling on a planetary trajectory. His distance from the central mass is constant, like that of the stationary observer, and he is equally incapable of communicating with the interior. This example shows that, with respect to event horizons, the free-falling observer does not share the situation of the inertial observer in a Minkowski universe in all respects.

Given the weakness of gravitational fields with which man has had any experience, event horizons have appeared to be primarily theoretical playthings without significance in practical astronomy. Yet event horizons may assume significance in two situation: the catastrophic collapse of very large masses; or the possibility of observations at cosmological distances, those approaching the dimensions of the universe at large.

PLATES

American Institute of Physics

Newton (engraving by W. T. Fry from the painting by Kneller)

AXIOMATA
SIVE
LEGES MOTUS

Lex. I.

Corpus omne perſeverare in ſtatu ſuo quieſcendi vel movendi uniformiter in directum, niſi quatenus a viribus impreſſis cogitur ſtatum illum mutare.

PRojectilia perſeverant in motibus ſuis niſi quatenus a reſiſtentia aeris retardantur & vi gravitatis impelluntur deorſum. Trochus, cujus partes cohærendo perpetuo retrahunt ſeſe a motibus rectilineis, non ceſſat rotari niſi quatenus ab aere retardatur. Majora autem Planetarum & Cometarum corpora motus ſuos & progreſſivos & circulares in ſpatiis minus reſiſtentibus factos conſervant diutius.

Lex. II.

Mutationem motus proportionalem eſſe vi motrici impreſſæ, & fieri ſecundum lineam rectam qua vis illa imprimitur.

Si vis aliqua motum quemvis generet, dupla duplum, tripla triplum generabit, ſive ſimul & ſemel, ſive gradatim & ſucceſſive impreſſa fuerit. Et hic motus quoniam in eandem ſemper plagam cum vi generatrice determinatur, ſi corpus antea movebatur, motui ejus vel conſpiranti additur, vel contrario ſubducitur, vel obliquo oblique adjicitur, & cum eo ſecundum utriuſq; determinationem componitur.

Lex. III.

American Institute of Physics

Newton submitted his Philosophiae Naturalis Principia Mathematica *to the Royal Society in three installments in 1686 and 1687. The reproduction here and on page 142 state the three laws of motion which were the foundations of mechanics for three centuries.*

AXIOMS, OR LAWS OF MOTION

LAW I

Every body perseveres in its state of rest, or of uniform motion in a right line, unless it is compelled to change that state by forces impressed thereon.

Projectiles persevere in their motions, so far as they are not retarded by the resistance of the air, or impelled downwards by the force of gravity. A top, whose parts by their cohesion are perpetually drawn aside from rectilinear motions, does not cease its rotation, otherwise than as it is retarded by the air. The greater bodies of the planets and comets, meeting with less resistance in more free spaces, preserve their motions both progressive and circular for a much longer time.

LAW II

The alteration of motion is ever proportional to the motive force impressed; and is made in the direction of the right line in which that force is impressed.

If any force generates a motion, a double force will generate double the motion, a triple force triple the motion, whether that force be impressed altogether and at once, or gradually and successively. And this motion (being always directed the same way with the generating force), if the body moved before, is added to or subducted from the former motion, according as they directly conspire with or are directly contrary to each other; or obliquely joined, when they are oblique, so as to produce a new motion compounded from the determination of both.

The three laws of motion as first translated into English by Andrew Motte, published in 1729, two years after Newton's death *(continued on page 143).*

Lex. III.

Actioni contrariam semper & æqualem esse reactionem : sive corporum duorum actiones in se mutuo semper esse æquales & in partes contrarias dirigi.

Quicquid premit vel trahit alterum, tantundem ab eo premitur vel trahitur. Siquis lapidem digito premit, premitur & hujus digitus a lapide. Si equus lapidem funi allegatum trahit, retrahetur etiam & equus æqualiter in lapidem: nam funis utrinq; distentus eodem relaxandi se conatu urgebit Equum versus lapidem, ac lapidem versus equum, tantumq; impediet progressum unius quantum promovet progressum alterius. Si corpus aliquod in corpus aliud impingens, motum ejus vi sua quomodocunq: mutaverit, idem quoque vicissim in motu proprio eandem mutationem in partem contrariam vi alterius (ob æqualitatem pressionis mutuæ) subibit. His actionibus æquales fiunt mutationes non velocitatum sed motuum, (scilicet in corporibus non aliunde impeditis :) Mutationes enim velocitatum, in contrarias itidem partes factæ, quia motus æqualiter mutantur, sunt corporibus reciproce proportionales.

LAW III

To every action there is always opposed an equal reaction: or the mutual actions of two bodies upon each other are always equal, and directed to contrary parts.

Whatever draws or presses another is as much drawn or pressed by that other. If you press a stone with your finger, the finger is also pressed by the stone. If a horse draws a stone tied to a rope, the horse (if I may so say) will be equally drawn back towards the stone: for the distended rope, by the same endeavour to relax or unbend itself, will draw the horse as much towards the stone, as it does the stone towards the horse, and will obstruct the progress of the one as much as it advances that of the other. If a body impinge upon another, and by its force change the motion of the other, that body also (because of the equality of the mutual pressure) will undergo an equal change, in its own motion, towards the contrary part. The changes made by these actions are equal, not in the velocities but in the motions of bodies; that is to say, if the bodies are not hindered by any other impediments. For, because the motions are equally changed, the changes of the velocities made towards contrary parts are reciprocally proportional to the bodies. This law takes place also in attractions, as will be proved in the next scholium.

Leverrier, who, in 1846, had predicted the existence of the planet Neptune from irregularities in the orbit of Uranus, in 1859 published new calculations of the orbit of Mercury. These showed an excess in the advance of its perihelion that could not be accounted for by the action of the other planets.

78 RECHERCHES ASTRONOMIQUES. — CHAPITRE XV.

Connaissant les valeurs de a et a', on peut éliminer $\delta\varepsilon$ entre les équations dont ces quantités sont les seconds membres. On tombe ainsi sur la relation

$$2{,}72\,\delta e + \delta\varpi = +\,10'',27.$$

Semblablement on tire, par l'élimination de δn entre les équations dont les seconds membres sont égaux à b et b',

$$2{,}72\,\bar{e}' + \varpi' = +\,0'',392.$$

On voit donc que la discussion des observations des passages de la planète sur le Soleil fournira une relation précise entre l'excentricité et la longitude du périhélie; mais que pour déterminer l'un de ces deux éléments, il sera indispensable de recourir à l'emploi des observations méridiennes.

Le mouvement annuel $2{,}72\,e' + \varpi' = +\,0'',392$ doit fixer notre attention; cette quantité étant essentiellement liée aux valeurs admises pour les masses des planètes. Les variations séculaires de l'excentricité et du périhélie de Mercure ont été calculées en attribuant aux masses des planètes les valeurs fournies par des considérations étrangères à la théorie de Mercure, mais qu'on avait lieu de croire fort exactes. On pouvait donc espérer que la discussion des observations de Mercure confirmerait simplement les recherches antérieures. Or il n'en est rien : nous voyons ici que le triple environ du mouvement séculaire de l'excentricité, ajouté au mouvement séculaire du périhélie, donnent une somme que les observations font plus grande de 39'' que celle qui résulte du calcul. La partie de cette somme, due à l'action de Vénus, est égale à 288'', par le calcul fondé sur la valeur 0,000 002 488 5 de la masse : et en conséquence, pour faire concorder la théorie avec les observations de Mercure, on devrait augmenter la masse, reçue pour Vénus, de près de *un septième* de sa valeur!

The page reproduced here, with the English translation, is taken from Leverrier's "Théorie et tables du mouvement de Mercure," Annales de l'Observatoire Impérial de Paris, V, 78.

Knowing the values of a and a', one may eliminate $\delta\epsilon$ from the equations of which these quantities form the respective second terms. One thus arrives at the relation:

$$2.72\ \delta\epsilon + \delta\omega = +10''.27\,.$$

Likewise, one derives, by the elimination of δn from the equations whose second terms equal b and b', respectively:

$$2.72e' + \omega = +0''.392\,.$$

Hence, one finds that the discussion of the observed transits of the planet across the sun furnishes a precise relationship between the eccentricity and the longitude of the perihelion, but that, in order to determine either of these two elements, one must fall back on meridian observations.

The yearly shift of $2.72e' + \omega' = +0''.392$ requires our attention; this quantity is tied essentially to the adopted values of the planetary masses. The variations per century of the eccentricity and of the perihelion of Mercury have been calculated on the basis of attributing to the planets masses furnished by considerations that lie outside the theory of Mercury but which one had reason to consider quite precise. Hence one might well have hoped that the discussion of the observations made on Mercury would simply confirm previous results. Now this is not so at all. We find here that roughly three times the (hundred-year) change in eccentricity, added to the shift of the perihelion, result in a sum that the observations render larger by 39″ than the value obtained from theoretical calculations. That part of the sum accounted for by the action of Venus amounts to 288″, a calculation based on the value for its mass of 0.000 002 488 5 (times the mass of the sun). Hence in order to make Mercury's theory agree with the observations, one should have to increase the mass of Venus by nearly one-seventh of its value!

Einstein, 1946

. . . Ich sende Ihnen einige Arbeiten.
Sie sehen, dass ich wieder einmal mein
Kartenhaus umgeworfen und ein neues
gebaut habe; wenigstens der Mittelbau
ist neu. Die Erklärung der empirisch
vollkommen gesicherten Perihel-Bewegung
des Merkur, macht mir grosse Freude,
nicht minder die Thatsache, dass
sich nun doch die allgemeine
Kovarianz des Gravitationsgesetzes
hat durchführen lassen.
Sie, Ihre Familie u. Koll. Smoluchowski
grüsst herzlich Ihr A. Einstein.

Estate of Albert Einstein

"I am sending you some of my papers. You will see that once more I have toppled my house of cards and built another; at least the middle structure is new. The explanation of the shift in Mercury's perihelion, which is empirically confirmed beyond a doubt, causes me great joy, but no less the fact that the general covariance of the law of gravitation has after all been carried to a successful conclusion."

By 1915, Einstein had developed his theory of gravitation to the point where he could explain Leverrier's findings. The material reproduced and translated here is from a letter, written December 15, 1915, to Einstein's friend and colleague, the Polish mathematician and physicist Wladyslaw Natanson (1864–1937).

Photograph by Lucien Aigner

Einstein, 1940, with his two co-workers, Valentin Bargmann (left) and the author (right), on their daily walk to the Institute for Advanced Study at Princeton.

Photograph by Ernest Bergmann

Fuld Hall, where Einstein's office was located.

Photograph by Lucien Aigner

At work in Einstein's office. The formula on the blackboard belongs to a unified field theory which Einstein, Bargmann (left), and the author right) were investigating.

Mount Wilson and Palomar Observatories

3 C 273 is one of the brightest QSOs. This photograph, which Sandage obtained on the Hale 200-inch reflector at Palomar Observatory, clearly shows a "jet," which extends roughly 150,000 light years from the core. The jet, a source of intense radio waves, consists of highly ionized gas, whose electrons spiral along the lines of a magnetic field. The material was ejected from the core with great violence at least a million years before the time of observation. (Light from 3 C 273 has been traveling several billion light years before reaching the earth.)

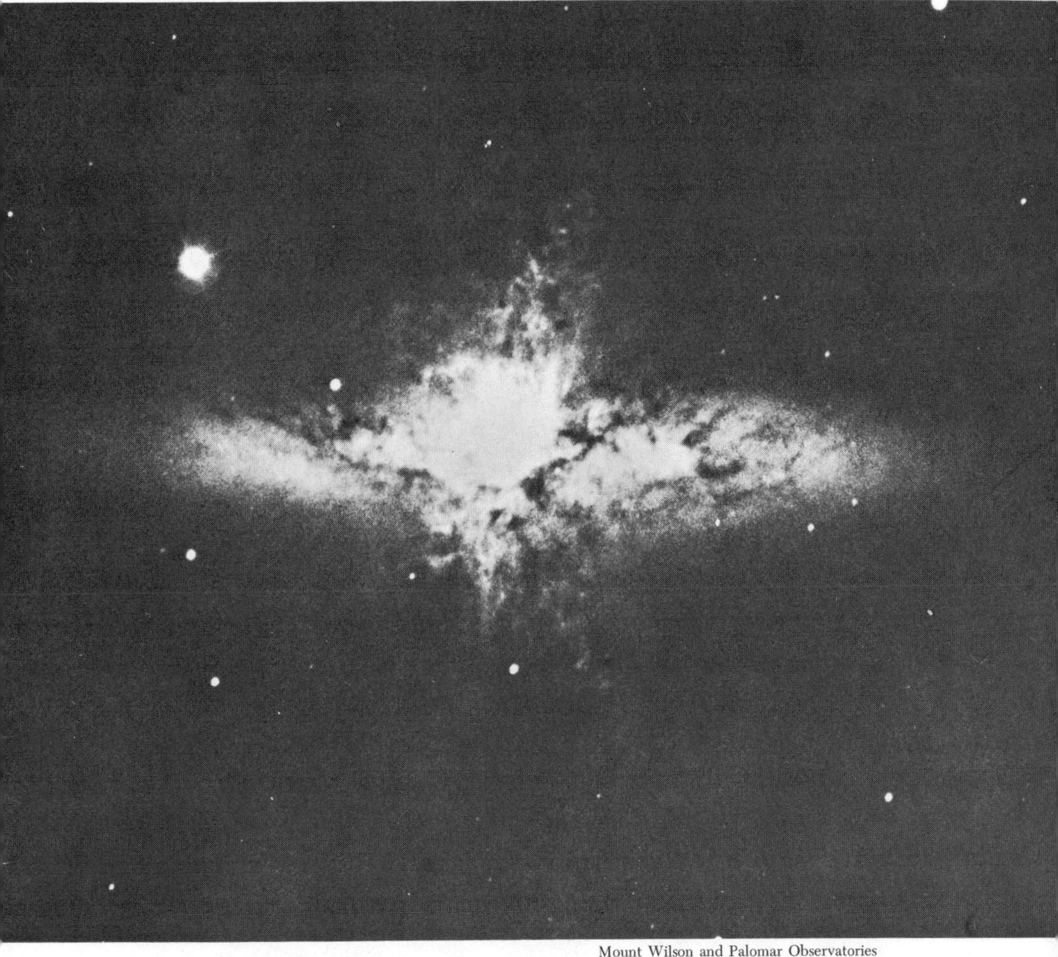

Mount Wilson and Palomar Observatories

Streams of ionized gas emitting strong radio signals occur in other systems that have undergone explosive change. This photograph, also by Sandage, is of M 82, a so-called radio galaxy. The exposure was made through a filter that passed only light from excited hydrogen gas in the Hα line. This technique brings out the bright filaments emerging on both sides of a rather flat galaxy viewed edge-on.

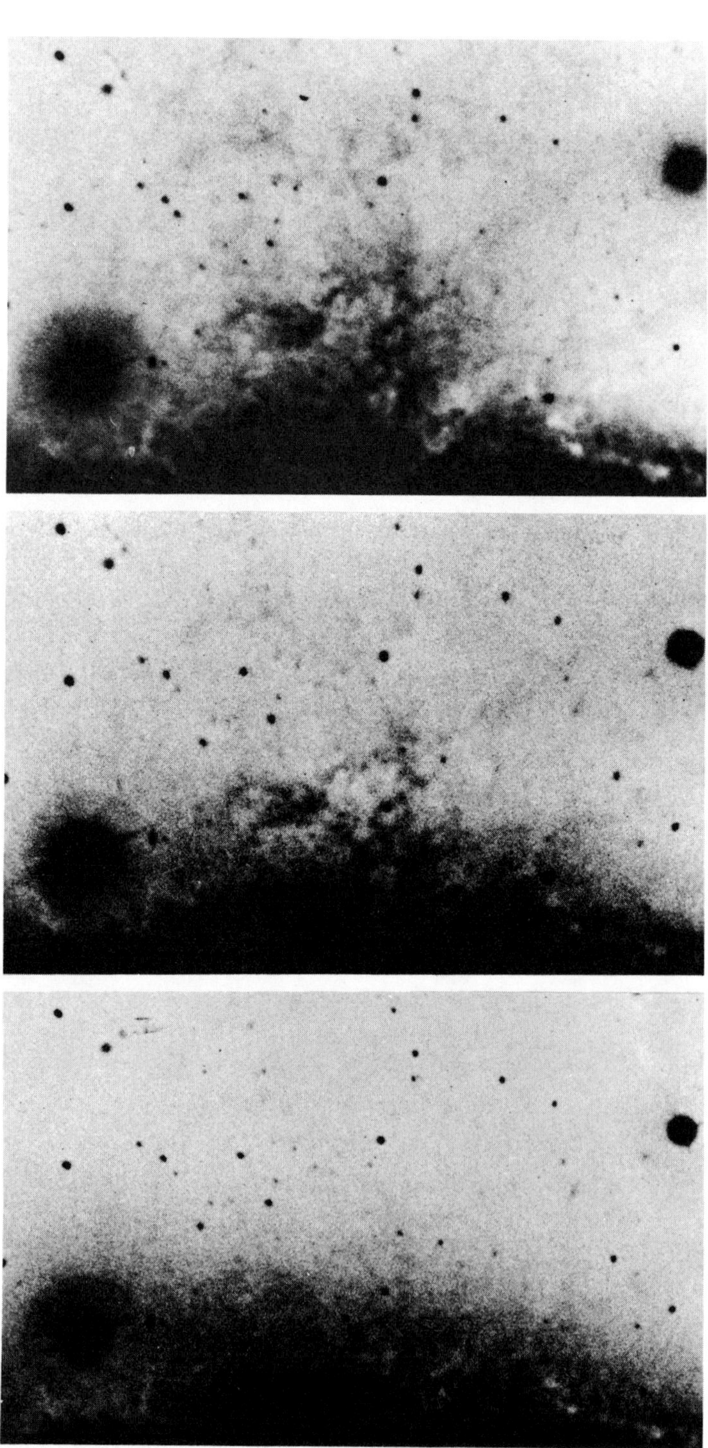

Mount Wilson and Palomar Observatories

These three photographs, which show a small segment of M 82 further enlarged, help elucidate the nature of the filaments. Reading from the top down, the first photograph was obtained through a yellow filter, the second through a polarizing filter aligned parallel to the plane of M 82, and the third through a polarizing filter at right angles to that plane. Whereas most of the filament structure shown in the first photograph remains visible in the second, almost none shows in the third. The high degree of polarization confirms that high-energy electrons spiraling in a magnetic field are responsible for most of the light issuing from the filaments. This evidence, together with stream velocities obtained spectroscopically, shows that the filaments at both sides were ejected from the disk of M 82 about one and a half million years before the time of observation.

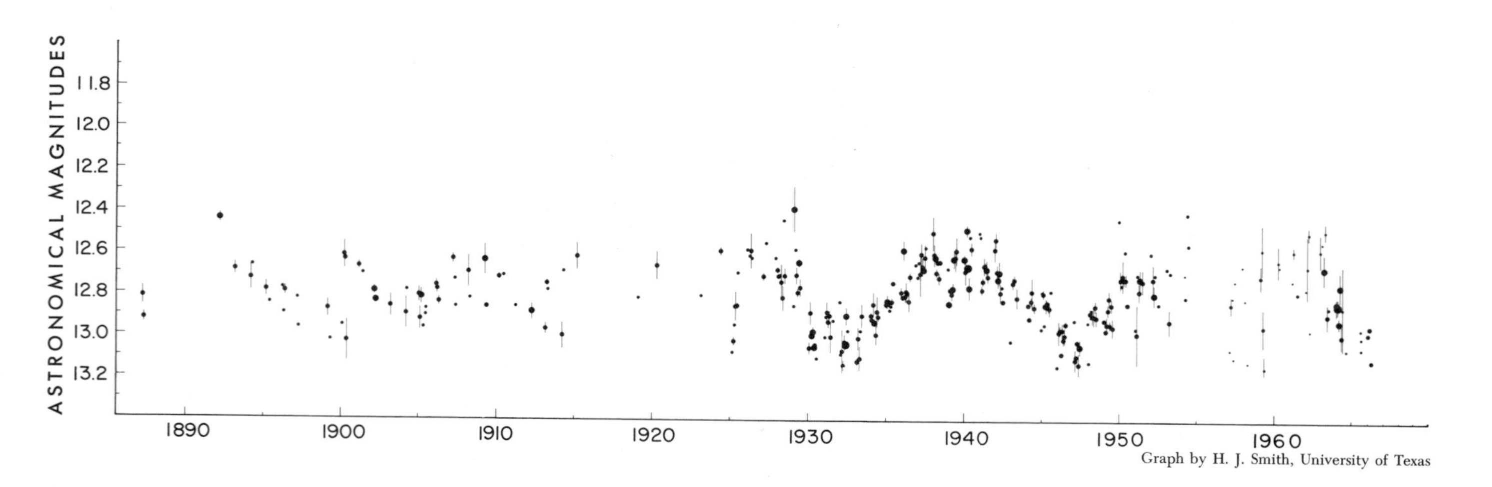

Graph by H. J. Smith, University of Texas

Jets and filaments tell of past explosions. Intensity variations observed in 3 C 273 and in other QSOs show that these objects have not settled down to a stationary state. This graph summarizes changes in the visual brightness of 3 C 273 which Harlan Smith and his coworkers discovered by reexamining photographic plates that had been exposed during an eighty-year period and that happened to include 3 C 273. The several thousand individual determinations of astronomical magnitude are plotted against the date of each picture. As one astronomical magnitude represents an intensity ratio of roughly 2.5:1, 3 C 273 has changed its light intensity by as much as 50 per cent several times during the period of observations.

III / RECENT DEVELOPMENTS

16 Gravitational Collapse

In 1939 Einstein published a paper in which he attempted to show that matter could not be so condensed that the Schwarzschild radius would fall outside the gravitating body and thus become a reality. In the same year, J. Robert Oppenheimer (1904–1967), then at the University of California, and two of his collaborators, claimed that any accumulation of matter that exceeded the mass of the sun by a fairly modest amount could not maintain a stable configuration but would continue to condense and shrink in size through its own gravitational attraction, terminating in cataclysmic collapse into its Schwarzschild radius. The issues of stability, progressive condensation, and collapse of very large masses are still with us. In 1963, astronomical objects, apparently with exceedingly large masses, were discovered by radio astronomy in conjunction with spectroscopic analysis; these objects have been given the name of *quasi-stellar objects* (*QSO's*) or *quasars*. This discovery has rekindled the interest of astronomers, cosmologists, and relativists in the problems attacked by Einstein and by Oppenheimer.

Oppenheimer and his collaborators postulated various types of condensed matter, for which they assumed specific laws tying the pressure to the degree of condensation. These laws, known as *equations of state*, appeared reasonable in the light of what was then known, or conjectured, about ultradense matter. Oppenheimer found that none of his assumptions resulted in a pressure sufficiently powerful to counteract the gravitational attraction with which a star of sufficiently large mass tends to pull in its own outer layers. According to the relativistic equations, as well as according to Newtonian theory, this gravitational attraction becomes more powerful as the surface approaches the center, but as density increases, pressure builds up as well. For objects of moderate size and mass, this pressure prevents the indefinite progress of gravitationally induced condensation. An equilibrium is eventually reached in which, at each distance from the center, the pressure is just large enough to keep in check the gravitational attraction, which includes the burden of all the outer layers. Once this equilibrium has been achieved, neither pressure nor density changes in the course of time, but both increase from zero at the surface of the star to their maximum values at the center.

If the total mass of an astronomical object exceeds a certain critical value, then, according to Oppenheimer and his collaborators, the increase of pressure with growing condensation does not suffice to bring about equilibrium, and collapse continues indefinitely.

Einstein's approach was different. He did not trust any specific assumption about the form of the equation of state, because he anticipated that, under sufficiently exotic conditions of density and pressure, the relativistic behavior of the constituent particles of matter might give rise to new phenomena which could invalidate theoretical results, such as those obtained by Oppenheimer. Einstein considered it safer, as long as no comprehensive theory of highly condensed matter had been worked out from first principles, to start from atomistic models certain to satisfy all requirements of relativity, however unrealistic or artificial the models might be in other respects. In order to avoid involvement in any hypotheses concerning the interaction of particles at very close range and at high energies, he built up a concentration of mass from particles traveling on circular plane trajectories about the common center. He assumed that these particles possessed such small individual masses that no two particles ever deflected each other. In this model, the distance of any one particle from the common center never varies, but particles are met with at all distances. If such an arrangement of atoms could be realized, the gross matter composed of such circulating atoms would exert no pressure in the direction of the radius (inward and outward), but considerable pressure at right angles to the radius. This cross-wise pressure in any layer would be related in a very specific manner to the mass density of that layer and to the total mass farther inside.

In building up mass concentrations in this manner, layer by layer, Einstein discovered that if he packed the particles sufficiently close there resulted a layer in which the individual particles must assume the speed of light in order to maintain their circular tracks, well before a Schwarzschild radius formed. A radially directed light signal emanating from the layer at which the particles moved at the speed *c* would succeed in reaching an observer stationed at any large finite distance from the common center, and its frequency would have undergone but a finite amount of red shift. It would take a similar light sight signal an infinite time to get out of the Schwarzschild radius, and its frequency would be red-shifted to zero. A light signal directed tangentially in the layer in which the particles reach the speed *c* would simply travel together with these particles and never leave that layer.

Since the 1939 papers by the Americans Oppenheimer and Hartland Snyder (1913–1962), and the Russian-born Canadian G. M. Volkoff, their work has been elaborated considerably. Because of the 1963 discovery of the quasi-stellar objects, whose masses are believed to be vastly larger than Oppenheimer's estimates of critical masses (which are in the neighborhood of the mass of our sun), various investigators have considered quite seriously whether there occur in nature concentrations of mass incapable of forming any equilibrium shape, or whether perhaps there are effects previously overlooked that might invalidate the arguments in favor of inescapable collapse.

Theorists have speculated whether matter, once compression reaches extreme values, might not build up enormous counter-pressures, so that eventually the collapse is halted. According to a careful analysis by John Archibald Wheeler and his collaborators at Princeton University,[1] this appears not to be the case if one limits the rise in pressure so that sound waves cannot spread in the compressed material at a speed exceeding that of light in empty space. These calculations were carried out on spherically symmetric bodies which did not revolve about some axis. Sufficiently rapid rotation might prevent collapse, but if too vigorous it would tear the quasi-stellar body asunder. Whether there are states of rotation that will maintain arbitrarily large masses stable remains to be settled.

No doubt, whether collapse is unavoidable or not, there are situations in which a sufficiently large mass is incapable of stable existence, where, in other words, there is neither rotation nor transverse pressure sufficient to keep the outer layers from falling into the interior. Such a mass will continue to contract, and even-

[1] Cf., for instance, John A. Wheeler, "The Superdense Star and the Critical Nucleon Number," in Hong-Yee Chiu and William F. Hoffman, eds., *Gravitation and Relativity* (New York: W. A. Benjamin, Inc., 1964).

tually an event horizon will form that will shield the core against any observation by an outsider who refuses to fall or jump into the star. Neither will an outside observer witness the closing of a door, as it were, against his inquisitive gaze. Rather, assuming continued transparency of the star, the outside observer will find that light signals emanating from the deep interior will become increasingly faint and red-shifted, with no well-defined endpoint. No matter how long he pursues the receding material with his telescope, it will continue to fade, but it will never vanish entirely from his view. The formation of the event horizon implies merely that, during his indefinitely extended observation, the outside observer can view only a finite portion of the history of the contracting core, and that he is precluded from ever receiving light signals leaving the core after the establishment of the event horizon.

17 Gravitational Radiation

Einstein's theory of gravitation predicts the existence of gravitational waves, with properties somewhat similar to those of electromagnetic radiation. Such waves are emitted by massive bodies undergoing acceleration; they propagate at the same speed as electromagnetic waves, at the speed of light. When the waves pass across massive particles, they cause them in turn to accelerate: after the passage of a pulse of gravitational radiation, each particle has its velocity changed. Moreover, when a gravitational wave pulse sweeps across a cloud of particles initially at rest relative to each other, they then will move relative to each other. These relative motions are perpendicular to the direction in which the gravitational wave traveled.

The general theory of relativity predicts the amounts of energy involved in the generation of gravitational waves. Compared to the generation of electromagnetic waves by moving electric charges, the intensities calculated for gravitational waves are exceedingly small; accordingly, prospects for their early experimental discovery are not good but programs toward this end have been under way since the late fifties.[1]

[1] J. Weber, *General Relativity and Gravitational Waves* (New York: Interscience Publishers, Inc., 1961).

Gravitational waves are *polarized;* they cause acceleration at right angles to the direction of propagation, resembling in this respect electromagnetic radiation. But electromagnetic waves and gravitational waves have quite different polarization characteristics. Suppose a gaseous cloud of electrically charged particles were suspended motionless in space, and then an electromagnetic wave were passed across it. All the particles should be driven to and fro in the same direction. This direction, together with the direction of propagation of the wave, forms the *plane of polarization.* The corresponding experiment with gravitational waves would involve the suspension of massive particles, again at rest at the outset. As the gravitational wave passes, the particles are set in motion relative to each other. If one particle be chosen as a standard of reference, particles on the left and on the right might swing away from each other, while particles above and below swing toward each other. At the next instant, all these motions are reversed: particles vertically above and below are moving outward; those on the sides, inward. Figure 60 shows how those particles initially forming a circle about the arbitrarily

Fig. 60. A Gravitational Wave

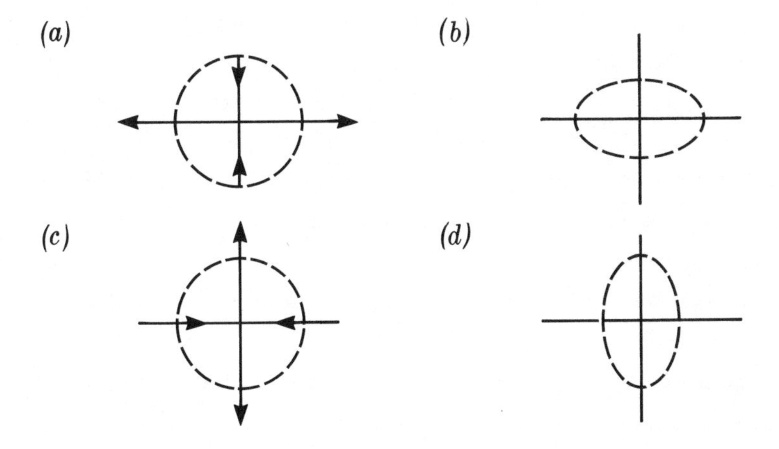

chosen reference particle will move at successive stages during the passage of a gravitational wave. (The wave is assumed to pass at right angles to the plane of the paper.) All these motions take place in the plane perpendicular to the direction of propagation of the gravitational wave.

The action of a gravitational wave may be compared with that of so-called *shear waves.* Shear waves represent one of the two types of acoustic waves met with in solid substances. The other type, *compression waves,* propagate faster than the shear waves. Both are observed in earthquakes. The difference in their respective arrival times at various seismic stations is the principal means for determining the location of the origin of a quake. Just as in shear waves, the state of polarization of a gravitational wave cannot be characterized by a single plane of polarization, but calls for two directions, both perpendicular to the direction of propagation and at right angles to each other, in which the motions of the participating particles are straight in and out.

Double stars constitute one kind of source of gravitational waves. A *double star* consists of two suns that orbit about each other, and about their common center of gravity, somewhat as the earth orbits about the sun. The whole motion takes place in a plane. This system, according to the theory, emits gravitational radiation in all directions. In the orbital plane, this radiation will have, as its planes of polarization, the orbital plane and the plane at right angles. A detection device placed far from the double star but in its orbital plane would receive the same polarized radiation if the whole double star were to be turned through an angle of 90° about the connecting line as an axis (Fig. 61). To obtain a differently polarized wave at the same location of the detector one would have to turn the double-star source of radiation through an angle of 45°.

The power of gravitational radiation can be estimated as follows: A double-star system possesses two kinds of energy. One

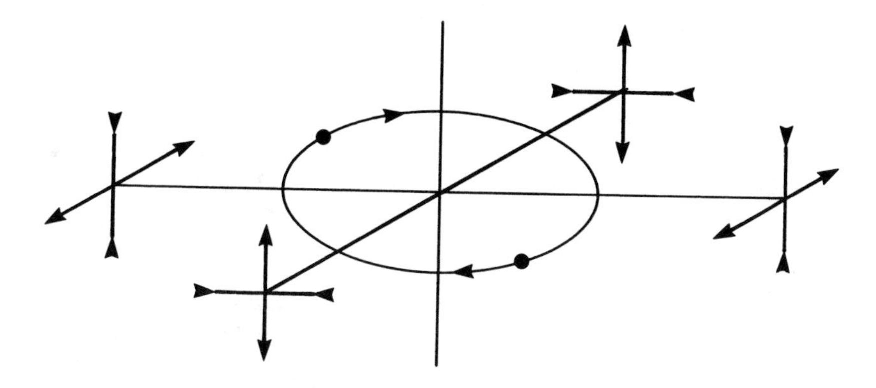

Fig. 61. Polarization of a Gravitational Wave Produced by a Double Star

is the rest energy, mc^2, residing in the total mass of the constituents. The other kind of energy is the mechanical energy recognized in Newtonian physics. The magnitude of the mechanical energy is of the order of mv^2, where v represents the velocity of one component relative to the other. The mechanical energy happens to be negative, because it requires work to pull a double-star system apart. At any rate, the ratio between mechanical and rest energy of a double star, (v^2/c^2), equals roughly the ratio between the Schwarzschild radii of the components and the distance between them (in these comparisons numerical factors, such as 2, $\frac{1}{2}$, are being disregarded). The time needed to dissipate an amount of energy comparable to the mechanical (binding) energy of the double star by its gravitational radiation will be the product of the time that it takes light to travel from one component to the other, multiplied by the cube of the ratio between the distance between the two stars and their Schwarzschild radii. If the time of dissipation be denoted by T, the distance between the two components by d, and the Schwarzschild radii by R, then

$$T \sim \frac{d}{c}\left(\frac{d}{R}\right)^3 .$$

The formula for the time required to dissipate the rest energy differs from the preceding expression in that the third power of (d/R) must be replaced by the fourth power.

Take a double-star system in which the masses are of the order of the sun and the distance between the components of the order of the distance sun-earth. In powers of 10, the value of (d/c) for such a system is of the order of 10^3 seconds, and the numerical ratio (d/R) of the order 10^8. Hence T comes out to an order of 10^{27} sec, or 10^{20} years. As the "age of the universe" is estimated to be only of the order of 10^{10} years, this rate of dissipation is negligible compared to the loss of energy through radiation of light and other processes involved in the aging of stars.

The rate of energy loss is increased tremendously as the characteristics of the double-star system are changed in the direction of more rapid revolutions. Without getting completely outside the realm of known stellar parameters, one might imagine a double-star system consisting of two white dwarfs. *White dwarfs* are stars with matter so condensed that they manage to compress the mass of a star, comparable to the mass of the sun, into a volume about the same as the earth. Accordingly, if two white dwarfs form a double star, their average distance from each other might be as little as 10,000 km, or 10^9 cm, and the ratio (d/R) might be of the order of 10^4. In such a system, the period of revolution drops to a fraction of a minute, and it takes but a fraction of a second for light to travel from one component to the other. The length of time in which energy of the order of the mechanical energy of the system is dissipated is now reduced to 10^{11} sec, or to 10,000 years; and that is a very short period in the history of an astronomical system. It takes but 10^8 years to radiate gravitationally an amount of energy corresponding to the rest energy of the double star, a period of time that is but a fraction of the periods usually quoted as the theoretical lifetimes of stellar objects. The rate at which the white dwarf system radiates gravitational power is 10^5 times the rate at which our sun radiates electromagnetic power.

Because white dwarfs are faint sources of light, it is not known how many white dwarfs are in a galaxy, and hence how likely it is that more than one component of a multiple-star system will be white dwarfs. Double stars containing one white dwarf are known. Perhaps the best-known example is Sirius, the dog star, whose other component is one of the brightest stars visible in the northern hemisphere. Even if several white dwarfs are combined into a system, they need not be so close to each other as to give rise to these extraordinary periods of revolution and to the corresponding tremendous rates of gravitational radiation. But white dwarfs are sufficiently small that such tight orbits are feasible without causing the components to collide with each other; yet they have large enough masses to be capable of giving rise to intense gravitational radiation.

18 The Search for Gravitational Waves

Systems of multiple stars generate gravitational waves, but ordinarily these waves carry energy that is insignificant compared to the radiant light energy emitted by incandescent stars. Besides, gravitational waves associated with multiple stars possess exceedingly low frequencies, frequencies directly related to the periods of revolution of these systems. In order to estimate the periods of revolution that might be realized in double stars, one may use a simple relationship: The period of revolution is of the order of the time that it takes light to cross the orbit of a double star, multiplied by the square root of the ratio (d/R). Consider once more the case of two white dwarfs orbiting about each other in close proximity. In that case, the period will be of the order of a fraction of a minute, hence the frequency between 10^{-1} and 10^{-2} cycles per second. Any periods observed in actual astronomical objects to date have been considerably larger, at least several hours (for satellites in our solar system) to several weeks (for double stars).

Multiple stars are, however, not the only astronomical objects that might give rise to gravitational waves. Gas clouds of electrically charged particles, called *plasmas*, are believed to be abundant in many objects that have undergone violent explosions or eruptions. The Crab nebula, as well as other remnants of *supernovae* (exploded stars) show luminous gas clouds which are probably plasmas. Radio galaxies and quasars almost certainly have large plasma clouds associated with them. Such plasmas are capable of oscillations, and these oscillations might give rise to gravitational, as well as to electromagnetic waves. No reliable estimates have been made of the intensities to be expected. Finally, the cataclysmic events likely to accompany gravitational collapse may touch off the emission of gravitational radiation (Appendix V). Some of these radiations may have frequencies considerably higher than those associated with orbiting stars. At present, however, such processes must be considered highly conjectural.

As for producing gravitational radiation in the laboratory, its intensity will depend both on the amount of mass set into oscillation and on the amplitude and frequency achieved. Joseph Weber at the University of Maryland has considered some schemes similar to those employed in the production of supersonic sound waves, though on a much larger scale; no actual device has been built.

How is one to detect the presence of gravitational waves? The acceleration of *one* test body alone cannot be observed. The principle of equivalence forbids setting up a local inertial frame of reference, with respect to which small bodies would be tossed to and fro by a passing gravitational wave; the same wave would have its effects on the free-falling frame of reference. Detection of gravitational waves must be based on the differential accelerations observed at distinct locations. A slight modification of the ideal detecting apparatus involving several small free-falling bodies some distance apart would involve one extended body in which the slightly different gravitational accelerations tugging at its parts cause internal stresses. Weber used this kind of arrangement, in the design of a gravitational wave detector. His sensitive element consists of a large block of aluminum fitted with strain gauges. The aluminum block is of a size that will resonate at a frequency of some 1500 cycles per second; in this frequency range, the detector is much more sensitive to any incoming gravitational waves than at other frequencies.

The sensitivity of the laboratory detector of gravitational waves is adversely affected by its relatively small size of but a few feet. At the resonant frequency, the wavelength of a gravitational wave is over a hundred miles; the difference in accelerations actuating different parts of the detector could be enhanced tremendously by an increase in size. Accordingly, it has been suggested that the whole earth be pressed into service as a detector of gravitational waves, incidentally of much lower frequency. From records of powerful earthquakes as well as from theoretical considerations, it is known that the earth as a whole resonates, with elastic vibrations, at a number of frequencies, of which the lowest correspond to periods of several minutes, up to half an hour. Only some of the earth's resonant modes of vibration could be activated by gravitational waves; the most symmetric modes, in which all parts of the earth move straight in and out in a sequence of compressions and dilations, cannot be excited by a stress which, by its very nature, requires that the response be a shearing motion (which alters the shape but not the volume). At any rate, it is possible to look for the excitation of shear modes by seismic and other detecting devices.

Finally, one might think of utilizing a satellite as one end of an observational base for differential accelerations, the other end being the earth. Though it is possible to detect very small changes in relative velocity by means of the Doppler shift of a frequency emitted by a radio transmitter on earth and rebroadcast by the satellite, other effects, such as the ordinary Newtonian attractions by other bodies in our solar system, will contaminate, and probably drown out, the signals caused by gravitational waves originating far from our solar system.

Currently, no gravitational waves have as yet been discovered, and there are few grounds for expecting early success in the search for them, though in 1967 Weber reported several observations on his detector that may indicate the incidence of gravitational waves.

19 Cosmology

From its inception, the general theory of relativity has stimulated cosmological investigations. *Cosmology* deals with the structure and the development of the whole universe. Like all other areas of scientific inquiry, it has both observational and theoretical aspects.

On the observational side, astronomers have been interested in the large-scale structure of the various observable astronomical objects. Astronomers have found that fixed stars are grouped in very large accumulations called galaxies. Our solar system is located in a galaxy consisting of some hundred million stars. Not all galaxies look alike; their differences probably represent, at least in part, different stages of development, leading perhaps from early accumulations of stellar material with relatively little structure to increasingly complex structures, with branches and convolutions, and containing increasing numbers of mature stars. Some galaxies provide evidence of major explosive occurrences, involving a large part of the galaxy. It is not known whether these galactic explosions represent stages in the development of most galaxies or whether they are very exceptional occurrences.

Galaxies form clusters, in the sense that the distances between galaxies within the cluster are much smaller than distances between clusters. So far galactic clusters are the largest-scale form of organization of matter that has been discovered. It looks as if clusters are distributed randomly throughout the universe.

Early cosmologies assumed that individual stars, and individual galaxies, go through life cycles possessing both beginning and end, but that the universe as a whole does not develop or age. According to these hypotheses, if an observer could survey the universe repeatedly, perhaps every few billion years, he would always observe the same population of stars and galaxies, with a roughly uniform distribution among the several types and ages of objects. Such a stationary over-all composition is, of course, possible only if there are mechanisms making possible both the emergence of new stars and galaxies, and the disappearance of objects that have reached the terminal stages of their development.

The early assumption of stationarity was abandoned when the American astronomer Edwin Powell Hubble (1889–1953), in the late twenties, established that the spectral lines of distant galaxies and nebulae all exhibit a shift toward long wavelengths, a red shift, which can be most naturally explained as the Doppler shift of bodies traveling rapidly away from us, and from each other. This interpretation is strengthened by the observation that out to very considerable distances, where estimates of optical brightness are still reasonably reliable, the red shift is proportional to the distance of the objects from us. This type of red shift is often referred to as the *Hubble effect,* or as the cosmological red shift.

If the Hubble effect is accepted as evidence that distant objects travel away from our galaxy, then all objects in the universe must be engaged in a process of dispersion, which leads to ever-diminishing density of matter in the universe. Conversely, by inference, the density of matter in the universe must have been much larger in the past than it is now. By following the motions of distant objects backward, one concludes that, if their velocity has remained the same as it is now for an indefinite period in the past, then the time when all presently observable matter was concentrated in a relatively small region was about twenty billion years (2×10^{10}) ago. This figure is often referred to as "the age of the universe." Its reciprocal, the rate of dispersion of matter at the present time, is known as the Hubble constant. Figure 62 shows the relationship between the recession of distant objects, limited by the slopes of the light cone *l.s.*, and the age of the universe.

Given the observed Hubble effect, it is not a foregone conclusion that the universe ever went through a cataclysmic period that can properly be considered its "birth." It is quite possible that the universe has always existed, and that it has always expanded at roughly the present rate, as indicated in Fig. 63. Or the universe might pass through alternate periods of expansion and contraction of sufficiently long duration so that no traces of the period of contraction preceding the present expansion could be discovered (Fig. 64). Such questions are being investigated by

Fig. 62. Linearly Expanding Universe

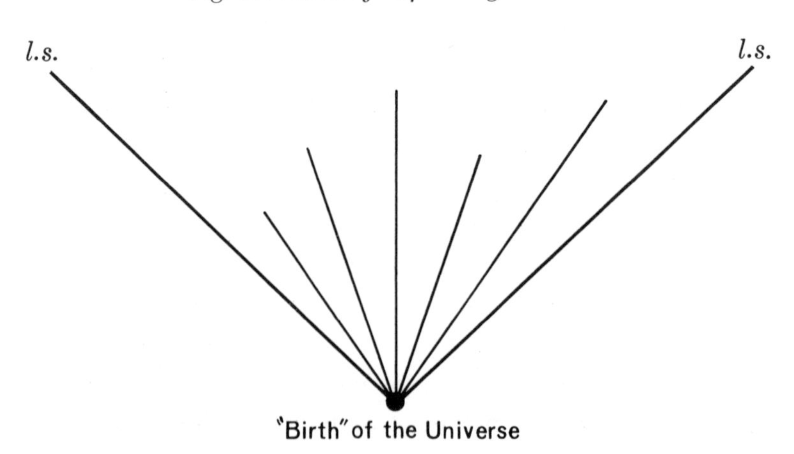

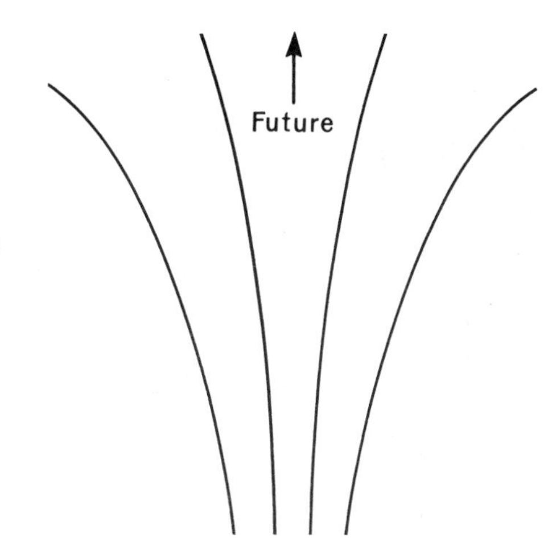

Fig. 63. Expanding Universe with Constant Hubble Coefficient

a combined observational and theoretical attack; definitive answers may not be forthcoming for a number of years.

A different approach has been proposed by the British astrophysicists Fred Hoyle, Hermann Bondi, and Thomas Gold.[1] They resurrected the principle that, as a whole, the universe should not change but always present the same over-all appearance. But accepting the Hubble effect as evidence that matter is constantly dispersing, they conjectured that there is present in nature a process in which matter is being created constantly at a rate just sufficient to offset the dispersion and to maintain the same average density of matter throughout the universe permanently (Fig. 65). Their hypothesis is known as "continuous creation," and also as the "steady-state hypothesis."

To explain why creation has never been observed is not too difficult. Assuming that creation takes place uniformly everywhere,

[1] H. Bondi and T. Gold, *Monthly Notices Roy. Astronom. Soc.,* **108** (1948), 252; F. Hoyle, *ibid.,* **108** (1948), 372. An excerpt of the former paper is to be found in Harlow Shapley, ed. *Source Book in Astronomy,* 1900–1950 (Cambridge, Mass: Harvard University Press, 1960).

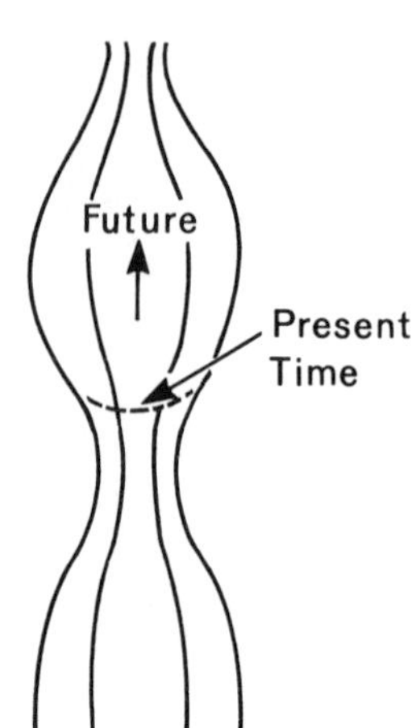

Fig. 64. Oscillating Universe

one needs to postulate that about one atom is being created every hour in a volume the size of a cube with an edge of one mile. In terms of observations sufficiently delicate to ascertain the coming into existence of a single atom, creation would thus be an exceedingly rare occurrence. Alternatively, one might conjecture that creation of matter takes place only in the interior of stars where temperatures, pressures, and densities are sufficiently different from those encountered on earth to provide conditions for physical processes not met with in our surroundings. If the creation of matter should be relegated to such exotic circumstances, the chances of observing the process would also be slim. Figure 65 illustrates the steady-state model by Hoyle, Bondi, and Gold. This model differs from the cosmological model illustrated in Fig. 63 (the so-called De Sitter model) only in that it assumes that continuous creation prevents a progressive diminution in the density of matter.

Astronomers have directed recent efforts toward contributing facts that might help to distinguish between different models of the universe, or perhaps to show that all those hitherto advanced are inadequate. If the past history of the universe involves periods in which the density of matter, or the distribution of galaxies of different types, was appreciably different from the present state of affairs, then one might see evidence of such earlier stages at very large distances, distances from which light has taken several billion years to reach earth. By looking sufficiently far out, astronomers might observe objects far back enough in time to represent an appreciable fraction of the age of the universe (Fig. 66).

Until recently, optical observations reaching this far out have been frustratingly sparse, for even the most powerful telescopes make visible only the very brightest objects at the distances of interest to cosmologists. Since 1963, interest has focused on classes of objects that are believed to be many times brighter than conventional galaxies, the quasars and the interlopers. It is to be hoped that in the years to come observations on these objects will provide significant cosmological evidence.

Yet another approach was initiated in 1965 by two groups, one at Bell Telephone Laboratories, the other at Princeton University. Assuming that the universe went through a period of very high

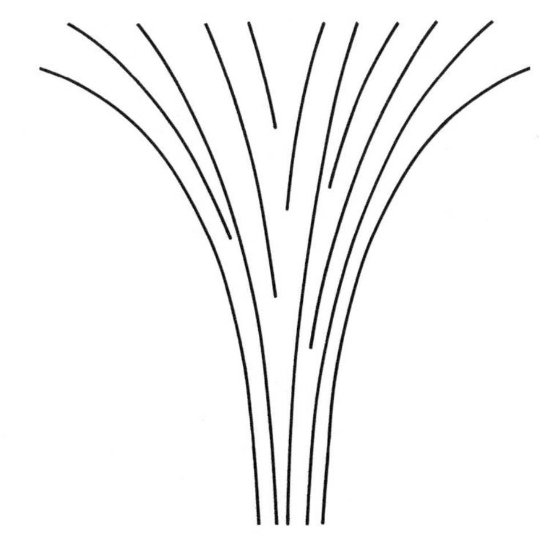

Fig. 65. The Steady-State Universe

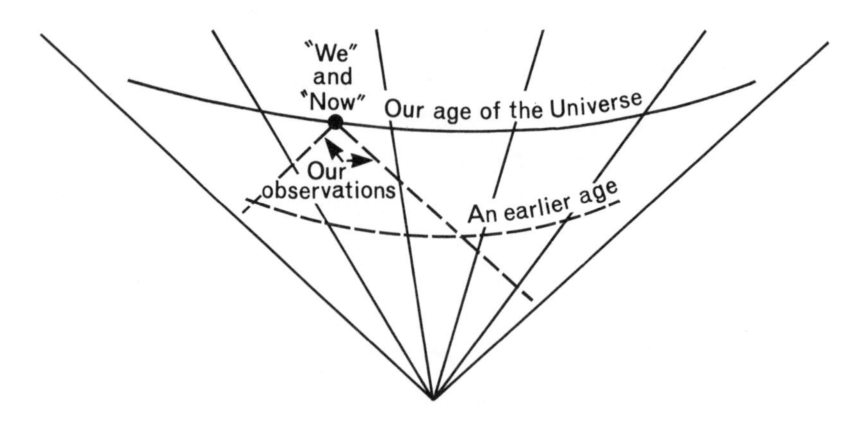

Fig. 66. Astronomical Observations Lead Billions of Years Back into the History of the Universe

matter density some 10^{10} years ago, the energy per particle during this period of high density must have been large. Because of the condensation of particles then prevailing, the particles present must have gone through frequent collisions with each other, giving rise to electromagnetic radiation of high energy and high intensity. If the universe expanded, and expanded rapidly in the early stages, the density of particles must have dropped quickly until a stage was reached in which the ambient radiation was no longer absorbed and modified by collisions with particles after passing through a small fraction of the universe. Beyond this stage, electromagnetic radiation must have continued to travel through the universe pretty much in the composition which it had at the stage when the particle density had dropped below a critical value. As electromagnetic radiation passed from the site of its origin to regions of the universe farther and farther removed, the continuing expansion was bound to result in an apparent red shift, analogous to the cosmological red shift discussed previously. This red shift would cause the radiation left over from the primordial high-density stage to appear today primarily as radio-frequency radiation.

For this reason, the Princeton group undertook a search for radiation of centimeter wavelength (so-called microwave radiation, lying in the same bands conventionally used for radar communications) not associated with the earth, the atmosphere, or specific celestial bodies, but that might come in evenly from all directions of the sky. This search turned out to be successful, in spite of very considerable experimental difficulties resulting from the low intensity of the phenomenon searched for. The radiation observed corresponds in intensity and wavelength distribution to the radiation emanating from a body at about 3.5° K (above the absolute zero of temperature), a bit cooler than liquid helium permitted to evaporate into the open atmosphere, and very much colder than liquid air. The discoveries of the group at Bell Telephone Laboratories, which actually preceded the work at Princeton, was incidental to the development of low-noise amplifiers. Their findings and those by Dicke's group at Princeton University confirm each other.

If these results are confirmed by further observations, the temperature of the matter that gave rise to the primordial radiation must be in the same ratio to the temperature observed now as the observed wavelengths are to those originally emitted. This ratio, in turn, equals the ratio between the age of the universe now and the age of the universe at the time that the density of particles in the expanding universe had just dropped to the critical value below which electromagnetic radiation would propagate freely through space. By substituting reasonable values for the critical density and by relating it to estimates of the present density of matter in the universe, one finds the temperature at the time of the primordial flash to have been in the billions of degrees, and the radiation itself in the nature of hard gamma radiation, more penetrating even than medical X rays.

If the observed low-level ambient radio frequency radiation really represents aged primordial "flash" radiation, then this evidence militates against the possibility of a steady-state cosmology. The existence of this radiation would certainly be compatible with the genesis of our universe through an initial convulsion ("big bang hypothesis") with subsequent expansion, but it would also not rule out the possibility of alternate expansions and contractions of the universe, provided only that before the present stage of expansion the universe had gone through a contraction sufficient to lead to a period in which high density of matter was accompanied by very high temperatures.

What does the general theory of relativity have to contribute to the problems of cosmology? There are surprising similarities between Newtonian and relativistic cosmology, but there is at least one major difference. If the space-time universe is not flat, then there is the possibility of its being finite without boundaries, just as the surface of a sphere is finite. In such a finite universe, one could travel in any direction at will without ever encountering a barrier marked "no trespassing." But if one traveled indefinitely in the same direction, one should one day find oneself back near the point of departure, just as a traveler would who traveled along the earth's equator going steadily east. A universe might be closed spacewise but infinite timewise, or vice versa. All these are possibilities permitted in a theory in which the geometry of space-time is not necessarily flat.

In the construction of cosmological models, it is usually assumed that at any given time all of the universe is filled with matter of the same average mass density and at the same average pressure, with both density and pressure permitted to change uniformly throughout the universe in the course of time. This assumption of *homogeneity* is often referred to as the *cosmological principle.* These averages of mass density and of pressure are to be understood to represent means obtained by considering space and time domains large enough to include many clusters of galaxies over a considerable period of time, though still small compared with the universe as a whole. The adoption of the cosmological principle for speculative purposes has no foundation in logic, or in the theory of relativity, but is based simply on the failure of all current observations to reveal any inhomogeneities on this space and time scale. It is entirely possible that the cosmological principle will be found not to represent the structure of the actual universe; this discovery would not shake the theoretical foundations on which cosmological hypotheses are based.

A further assumption, called *isotropy*, which also has a purely empirical basis, is concerned with the essential equivalence of all spatial directions. Though many individual galaxies are not spherical in shape but flattened, somewhat like lenses, and though spiraling branches provide evidence of rotation about some axis, there does not appear to exist in the universe as a whole any preferred plane, or any axis of rotation. Hence most cosmological models are *isotropic* as well as *homogeneous*. These assumptions refer to three-dimensional isotropy and homogeneity. There is always a well-marked time axis which represents the mean local state of motion of matter, and in most cosmological models there are three-dimensional spacelike "surfaces" that are everywhere perpendicular to the local time axis. Each such surface (Fig. 66) represents one instant of "time" on the cosmological scale. Accordingly, cosmological models never have the full symmetry suggested either by Lorentz invariance or by general covariance. All cosmological models that can be conjured by a fertile imagination have preferred frames of reference, albeit somewhat fuzzily defined.

That this is so does not represent an internal contradiction in the concept of a relativistic theory of the universe. The requirements of relativistic invariance concern the laws that must be obeyed by all models of the universe; each individual model has its own preferred time axis, and it may well possess additional features that tend to single out preferred frames of reference. There are, incidentally, cosmological models that are homogeneous but not isotropic; one of the best-known was constructed by the mathematical logician Kurt Gödel at Princeton. In the Gödel universe there is built in everywhere a sense of rotation.

Return once more to a cosmological model that is three-dimensionally homogeneous as well as isotropic. The average matter filling such a universe model can, at each instant of time, be represented in terms of just two parameters, its mean mass density and its mean pressure. The curvature tensor depends on only two components that are not algebraically related to each other by symmetry requirements alone, and these both form part of Einstein's tensor. Weyl's tensor vanishes. As for the spatial part of the curvature, there are three possibilities: that it be positive, zero, or negative. If it be positive, the universe would be spatially finite; in the two other cases it would be spatially infinite.

Assuming a positive curvature, in which the universe as a whole possesses a finite volume, average pressure and mass density are capable of changing in the course of time. Because the laws of general relativity require that Einstein's tensor be proportional to the energy-stress tensor (and pressure forms part of the stress), changes in mass density and pressure bring about changes in the spatial curvature, and hence in the total volume, in the course of cosmological time. In this fashion, it is possible for matter even in a finite universe to stream together or apart, to condense or to undergo rarefaction, without ceasing to fill the available space of the universe, without forming large-scale lumps or leaving voids.

To repeat once more, the general theory of relativity neither requires the cosmological principle, or any of its variants, nor does it favor any one among the many different cosmological models that have been constructed in accordance with the requirements of the cosmological principle. The so-called steady-state model of Hoyle, Bondi, and Gold violates the laws of general relativity, because creation of matter out of nothing violates the conservation laws of energy and linear momentum, which form an integral part of the theory as we know it. This contradiction can be remedied formally if one hypothesizes the simultaneous creation of particles of negative mass endowed with such properties as to defy most ordinary methods of detection. It is, of course, possible that eventually such particles will be discovered to exist, although presently there is not the slightest evidence to that effect. Whether the laws of general relativity hold on the cosmological scale or whether new phenomena presently unknown have a major effect on the large-scale structure of the universe can be decided neither by the operation of "pure reason" nor by a crassly empiricist approach, but only by inspired observational and theoretical approaches reinforcing each other.

20 Current Observational Programs

Much current laboratory research in general relativity is directed toward confirmation (or disproof) of Einstein's theory. As the theory is predicated on the perfect equality of inertial and gravitational mass, Dicke's recent improvement of Eötvös' original confirmation of this equality has secured the foundation of general relativity to an accuracy of one part in 10^{11}. Other predicted effects of the theory have been examined with varying degrees of success.

As mentioned earlier, lines of atomic spectra that originate near a large mass are shifted toward the red (long wave) end of the spectrum. This so-called gravitational red shift was first observed in the spectrum of a white dwarf, Sirius B, a star consisting of very dense matter, on whose surface the gravitational field is much stronger than on the surface of any member of the solar system, including the sun itself. As the radius of Sirius B cannot be determined very accurately, this observation represented only a rough test of the theoretical prediction. The gravitational red shift in the spectrum of the sun is only about one-thirtieth that of Sirius B; other factors cause shifting and broadening of the sun's spectral lines. For some time, the difference between the shifts observed in the center of the sun's disk and those observed near its rim was considered a serious objection to the validity of general relativity. According to current opinion, this differential effect is probably related to the convective currents within the visible layers of the sun; continuing observations are aimed not so much at testing general relativity but at better understanding of the constitution and the dynamics of the sun. The laboratory experiments of Pound demonstrated agreement between predicted and observed red shift within the limits of experimental accuracy, approximately 1 per cent; hence the gravitational red shift must now be considered confirmed with gratifying precision.

Leonard L. Schiff of Stanford University and others have emphasized that gravitational red shift would agree not only with general relativity but also with any other theory in which loss of potential energy of light traveling against the pull of a gravitational field leads to a corresponding change in frequency. That frequency and energy of light be proportional has formed one of the main assertions of quantum theory ever since the pioneering work of the German physicist Max Planck (1858–1947) in 1900. In reply to Schiff's contention, it must be admitted that many other conceivable theories of gravitation incorporate gravitational red shift, but there are also possible theoretical constructions that do not; to this extent, the experimental confirmation of the gravitational red shift is not without interest.

The bending of light rays passing near a large mass has been observed during total eclipses of the sun. Observations that must be performed in out-of-the-way locations and during periods occurring only every few years and lasting but a few minutes cannot be refined and repeated as can more conventional astronomical observations; the accuracy of the results always leaves something to be desired. For this reason, several groups of investigators, among them Henry Allen Hill at Wesleyan University, are working on an approach that will permit making observations in broad daylight. Briefly, by scanning the immediate vicinity of the sun with techniques similar to those employed in making television pictures, one can recognize stars close to the limb of the sun by the minute increase in the brightness of the sky, an increase too small to be seen by the unaided human eye. If these scanning techniques can be perfected so as to yield locations with an accuracy of a fraction of a second of arc, data will be obtainable even when the sun is not eclipsed.

Bending of light rays is to be considered a direct consequence of the curvature of space-time. The satellite experiments mentioned in Chapter 11 are concerned with another aspect of space-time curvature; hence these two types of proposed experimental programs complement each other.

Yet another experimental program bearing on space-time curvature has been proposed by Irwin Ira Shapiro of Massachusetts Institute of Technology. Light passing near the limb of the sun will not only be bent but also delayed in transit, because of distortions in both space and time distances. This delay cannot be measured by ordinary optical means, because the instant at which light leaves any object appearing close to the sun is unknown. Modern radar techniques have, however, succeeded both in sending radio pulses to the moon, and even to Venus and Mercury, and in recognizing the reflected signals. Shapiro has proposed to send radar signals to Mercury and Venus when they are beyond the sun and close to it in the line of sight (called *conjunction*) and to time the return signals with an accuracy that would reveal the increase in transit time; this accuracy is well within the reach of present techniques.

The advance of the perihelion of Mercury was considered a settled matter after 1915 when Einstein derived from his theory a definite expression, which was then discovered to agree very accurately with an effect known to astronomers specializing in the calculation of planetary orbits. Planetary ellipses all tend to change their orientations in space because of the perturbations caused by neighboring planets and by Jupiter. The orbit of Mercury, in particular, is subject to perturbations caused primarily by the neighboring planet, Venus, and by Jupiter, next to the sun the largest body within the solar system. Mercury's orbit turns at a rate of approximately one degree of arc per century. The theory of planetary perturbations, developed to perfection in the course of the nineteenth century primarily by the French school of mathematicians, Pierre Simonde Laplace (1749–1827), Joseph Louis Lagrange (1736–1813), Urbain Leverrier (1811–1877), and others, accounted for all but 1 per cent of the observed perihelion advance, and it was this tiny discrepancy, forty-three minutes of arc per century, that general relativity was able to account for.

The excellence of this agreement has been called in question by Dicke. Dicke had become interested in a modification of Einstein's theory of gravitation, presently known as the *scalar-tensor theory*, which had been developed independently by Einstein and Peter G. Bergmann, the German physicist Pascual Jordan, and Y. R. Thiry, a Frenchman. This speculative development was based originally on an attempt by Theodor Kaluza (1886–1954) in Germany to include the electromagnetic field in the geometry of space-time by postulating a fifth (unobservable) dimension. The scalar-tensor theory modifies Kaluza's attempt by adding to the gravitational and electromagnetic fields one extra variable quantity. Jordan attempted to turn this extra mathematical quantity to advantage in cosmology by tying it to a proposal made in 1937 by the British physicist Paul A. M. Dirac.

Dirac had noted that certain very large numbers in physics and cosmology tend to be of the orders of 10^{20}, 10^{40}, and 10^{80}. For instance, the ratio of electric and gravitational forces that two electrons exert on each other is, very roughly, 10^{40}, independently of the distance between the electrons. As another example, an estimate of the total number of protons and neutrons in the universe, whose size is estimated from Hubble's cosmological red shift, yields a number of the order of 10^{80}. A complete theory of the physical universe should account for these numerical values, a seemingly hopeless task. Dirac suggested that this formidable difficulty would not arise if all these large numbers were not true constants but slowly variable, so that their present values were characteristic for but one instant in the development of the cosmos. All of them would be, more or less, powers of one among them, of the order of 10^{20}, hence their falling into those conspicuous ranges that are separated from each other by factors 10^{20}.

Jordan conjectured that the available extra variable of the scalar-tensor theory might be identical with Dirac's speculative cosmological variable, and Dicke shared this view. Accordingly, both undertook to search for effects that would result from slow but persistent changes in the characteristics of elementary particles, and from other aspects of the altered field equations. The new theory also obeys the principle of general covariance; hence no changes are to be expected in those effects that are explained primarily in terms of the principle of equivalence.

According to Dicke, the scalar-tensor theory leads to a slightly different rate of perihelion advance than Einstein's original theory, the difference amounting to about one-tenth of Einstein's value. For this reason, Dicke reviewed all the possible effects that might contribute to Mercury's perihelion advance. He discovered one that had not been considered previously, the possible oblateness of the sun. Newton developed his theory of planetary orbits on the assumption that both sun and planet are spheres, as the gravitational effects of a sphere are the same as those of a point mass. A significant deviation from spherical shape would perturb the planetary orbit and lead to a perihelion advance.

By scanning the sun's edge and continuing taking data through several months during the summer of 1966, Dicke endeavored to eliminate from his final evaluation such effects as the refraction of the earth's atmosphere, which produces an apparent oblateness of the sun (and also of the moon) that is visible to the unaided eye when the sun is low in the sky. As Dicke was looking for a flattening of less then 10^{-4} (one ten-thousandth) of the sun's radius, elimination of masking effects presented a major challenge; Dicke believes that he has in fact discovered a flattening of the expected magnitude, and he concludes that instead of forty-three seconds of arc per century the remaining perihelion advance to be explained by the post-Newtonian theory amounts to only thirty-nine seconds of arc.

The determination of the sun's shape to an accuracy better than one part in ten thousand is so difficult that Dicke's first result will be tested by the taking of further data, and Dicke himself is planning to do so. The exact amount of the discrepancy between the observed perihelion advance of Mercury and the value accounted for by the perturbing effects of other planets will also be calculated anew by a group at Massachusetts Institute of Technology, which will combine the best available data on planetary orbits obtained by conventional optical methods with data secured by the new techniques of interplanetary radar.

The elaboration of the scalar-tensor theory, and perhaps of other similar speculative structures, has not been completed. Whatever the results of further observations, their bearing on the competition between Einstein's original theory of gravitation and the new variants is not yet entirely clarified. Wholly apart from these questions, though, successful high-precision determinations of the sun's shape will contribute profoundly to an understanding of its internal structure. As the sun's interior is believed to be fluid, an oblate shape would be almost certain proof of a high rate of rotation of that interior, perhaps more than ten times faster than the visible surface.

This concludes a very brief account of some current observational programs concerned with confirming or refuting the general theory of relativity. An entirely different connection between gravitational theory and observations appeared possible when the discovery of quasi-stellar objects was announced in 1963. These incredibly luminous objects combined copious emission of electromagnetic radiation of radio wavelength with very large frequency shifts toward the red, ranging among the specimens discovered so far from a little over 10 per cent to well above 100 per cent. It was also discovered, by Harlan Smith (now at the University of Texas) and E. Dorritt Hoffleit, Director of the Maria Mitchell Observatory, that some of the QSOs had highly variable luminosity, with rather ill-defined periodicities of a few years, punctuated occasionally by sudden changes occurring within the space of a few days or weeks.

Immediately on the discovery of the QSOs, every effort was made to decide whether the observed large red shifts were gravitational or cosmological (Hubble effect), because the estimate of their distance from earth depended thereon. The decision then, and the majority opinion at present, was that the red shifts were cosmological, because no reasonable model seemed capable of explaining a large gravitational red shift in combination with the other information on luminosity and approximate surface temperature. If this interpretation can be maintained, then QSOs are placed at distances of 10^9–10^{10} light years from earth. This would imply that a careful study of their distribution throughout the sky might provide significant information on the large-scale structure of the universe. No other objects known to astronomers have an intrinsic luminosity comparable to that of QSOs, according to this view; for this reason alone, no other class of objects, not even large galaxies, can furnish the same kind of information. In this context, cosmologists often speak of *standard candles,* any class of identifiable astronomical objects of large luminosity which are, one hopes, distributed uniformly and at random throughout the universe. The brighter, and hence the more highly visible, each individual standard candle is, the more useful the data that may be inferred from statistical information about its apparent distance, luminosity, and red-shift characteristics.

In 1966, Halton C. Arp of the Palomar Observatory published information indicating that an improbably large percentage of QSOs was to be found in apparently close proximity to other astronomical objects whose distance from earth was known to be of the order of 10^8 light years. Arp's data are statistical, in the sense that they suggest actual association of the QSOs with other objects not because of apparent proximity in individual cases, but because of the preponderance of such cases. The argument is subtle, and it is not yet resolved. If it should prevail, then all present estimates of luminosity, mass, size, and other characteristics of QSOs would be subject to being scaled down several orders of magnitude. The observed red shifts might then be gravitational, indicative of very exotic structure, or they might be interpreted as Doppler shifts. The latter interpretation would imply that all known QSOs are receding from earth at relativistic speeds, an implication hard to accept.

If the observed red shifts should be gravitational, then QSOs would give rise to the largest gravitational force fields observed anywhere in the universe. They would become outstanding candidates for the relativists' laboratories-away-from-earth, though a bit hard to get to. Obviously, QSOs are among the least-understood astronomical objects known; whatever the ultimate decision on their distances from earth, they will continue to fascinate the relativist.

21 Particle Motion

Traditionally, relativists have been concerned with large and concentrated masses, gravitational waves, and cosmological problems. In these fields and through observational tests of the theory itself, the general theory of relativity comes into closest contact with experiment and observation. There has also been much further analysis of the theory itself, which differs in a number of important respects from all other physical theories. The remaining sections of this book are concerned with theoretical work, some of it still under way.

The chief characteristic of the general theory of relativity remains, of course, that its laws are the same whatever coordinate system is chosen. In empty space, the gravitational field obeys field laws that in some respect resemble the laws governing other physical fields, foremost among them the laws of electricity and magnetism. But whereas all the other field laws known in physics look different in form, depending on the type of coordinates employed (linear or curvilinear, rectangular or oblique), Einstein's laws of the gravitational field take exactly the same mathematical form no matter what coordinate system is chosen. This freedom of choice can be exploited in a number of different ways. A particular technique consists of comparing the different forms that a particular field will take when described in terms of different coordinate systems, and of utilizing the fact that all these forms must obey the same laws (differential equations). Two coordinate systems may be identical with each other for some range of the time coordinate but differ outside that range (Fig. 67). The corresponding two forms that a particular field will take will also be identical within the range in which the two coordinate systems employed coincide but will differ from each other outside that range. It follows that in general relativity the mathematical form of the field within a chosen period of time cannot uniquely determine the continuation beyond that range.

This lack of uniqueness in the continuation of solutions of the field laws along the time axis (or, for that matter, along any other axis) has profound implications for their mathematical structure. Superficially, the laws of the gravitational field (as postulated by general relativity) resemble those of any other field theory. The number of equations equals the number of components of the

Fig. 67. Two Coordinate Systems that Coincide for Part of the Time

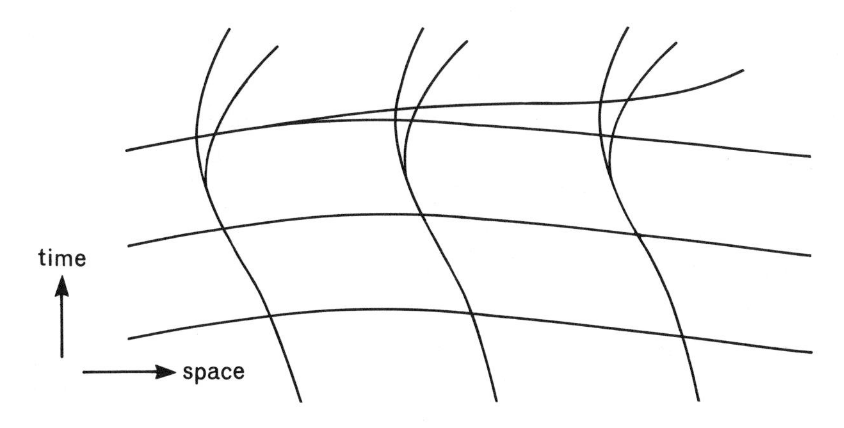

field, ten. But in previous field theories each of these equations fixes the time rate of change of one of the field variables: If at one time the values of all field variables at every space point are determined experimentally, then the field equations determine the rates at which all of these variables change, and thereby their values at all other times.

In general relativity the four coordinates used to identify world points may be continued in an arbitrary fashion from a chosen starting time, and the continuation of the field into the future (or into the past) is, to that extent, arbitrary. Accordingly, of the ten field equations only six bear directly on the continuation into the future, leaving four of the field variables free to assume arbitrary rates of change. The remaining four field equations set up conditions restricting the choice of field variables at the chosen starting time.

These restrictions are without parallel in ordinary field theory. They are, however, characteristic of any theory that matches the number of equations to the number of field variables while obeying the principle of general covariance. If, for instance, the general theory of relativity is enriched by the inclusion of the electromagnetic and other fields without destroying its general covariance, four of the equations will still have no bearing on continuation but will represent restrictions on the permissible initial situation.

If the restrictions on the initial situation have been satisfied at one time, the field equations automatically provide for their being preserved; they need not be imposed anew at other times. To this extent, the four field equations providing the initial restrictions and the six field equations governing the continuation in time harmonize with each other. Together they govern the motion of particles within the field.

In a typical situation in celestial mechanics, most of space is empty of matter and contains only the gravitational field which obeys the field equations of the vacuum. The celestial bodies, the sources of the field, occupy small regions separated from each other by vast voids. Are these sources free to move about arbitrarily or are they constrained by the surrounding field? In Einstein's theory of gravitation constraints do in fact exist, whereas in other field theories, the sources are free to move in any way. In electrodynamics, for instance, the sources of the electromagnetic field, the electric charges, may be guided and moved about arbitrarily through the application of non-electric forces, and the surrounding electromagnetic field can still be made to satisfy Maxwell's field equations. In gravitational theory, however, no field can be constructed mathematically in the space surrounding the gravitating masses obeying Einstein's equations unless each of the masses satisfies four requirements. One of these conditions determines the time dependence of the mass of the ponderable body; the remaining three govern the three components of its motion. If only one ponderable body is present, these four restrictions specify that its mass, as well as its velocity, remain constant. In the presence of other bodies, the same conditions will bring about the mutual attraction of ponderable masses (their accelerations toward each other).

At first sight, this theoretical result appears to lead to difficulties. General relativity has no need for separate and distinct laws for the field and for the motions of ponderable bodies, because the latter are already contained in the former. Hence the theory rules out any possibility of manipulating the sources of the gravitational field by external means in a manner analogous to the manipulation of electric charges, which is permitted by the theory of the electromagnetic field, and is readily performed in the laboratory. Does not the greater rigidity of the gravitational theory contradict everyday experience?

The answer is inherent in the principle of equivalence. To manipulate either electric charges or gravitating masses one needs tools. Whereas there are tools that are electrically inert whose presence contributes nothing to the electric field, tools capable of moving about ponderable bodies must themselves possess such properties as mass and stress, which in turn cause the tools to be sources of the gravitational field. Even before relativity, Newton's third law of mechanics stipulated that any agency that was capable of exerting a force sustained a counterforce. Depending on its own mass, the tool itself undergoes acceleration. When the Newtonian theory is modified and replaced by Einstein's theory, it is found that the total gravitational field obeys all the conditions imposed by the theory only if the law of action and reaction is satisfied. Thus the threatened paradox does not in fact arise. On the contrary, since the field laws imply the laws governing the motions of the sources of the field, a mathematical harmony between these sets of laws is assured. Assured, too, is the conservation of total mass and of total linear momentum, which other theories often must achieve by considerable mathematical effort.

22 Quantum Theory of Gravitation

Twentieth-century theoretical physics is characterized by the emergence of two conceptual structures, relativity and quantum physics. Quantum theory antedates relativity by five years; Planck published his first paper on the quantum theory of black-body radiation in 1900. This early paper was not concerned with the quantum physics of atoms. It proposed to explain the spectral distribution of electromagnetic radiation given off by heated bodies. Planck conjectured that the energy in electromagnetic radiation occurs only in lumps whose size is proportional to the frequency of the radiation emitted. Without this assumption one cannot understand why a heated body does not give off radiation of arbitrarily high frequencies at rates that cause the total amount of radiant energy to be infinite. Thus the birth of quantum theory was associated with properties of the electromagnetic field. Thirteen years later Niels Bohr (1885–1962), the Danish physicist, extended the quantum concept to atoms and to atomic spectra.

Initially quantum theory was concerned primarily with the

occurrence of energy in lumps. Gradually it was realized that this lumping of energy, which experimentally is beyond doubt, cannot be reconciled with classical notions of the behavior of particles. The German physicist Werner Heisenberg argued in 1927 that *in principle* one cannot assign to a particle both an exact location and an exact state of motion at the same time, nor perform experiments and observations determining both these quantities to an arbitrary degree of accuracy. According to Heisenberg, the product of the inaccuracies incurred in the determination of location and linear momentum can never be less than approximately Planck's constant, a universal physical quantity whose value is 6.6×10^{-27} erg sec. This limitation, known as the *uncertainty principle,* permits the establishment of a theoretical structure in which certain quantities, such as the energy and the angular momentum, are capable of only discrete values even though the ordinary laws of mechanics retain their validity, at least for the averages of all physical quantities. It is characteristic of quantum theory that the outcome of most measurements cannot be predicted with complete certainty on the basis of past experience. At best, probabilities can be assigned to the various possible values. In other words, quantum theory does not predict the result of a single measurement. It does tell how frequently each possible outcome will occur if the same experiment and measurement are performed a large number of times.

The uncertainties required by Heisenberg's principle are quite sizable in terms of atomic phenomena. They are, however, usually insignificant when bodies of ordinary mass and size are involved; hence the uncertainty principle causes no difficulties in the interpretation of everyday experience.

As far as physicists currently understand these matters, the fundamental ideas of quantum physics and the uncertainty principle, in particular, apply universally to all physical phenomena.

In fact it would be quite difficult to imagine interactions between two physical systems one of which is subject to the uncertainty principle whereas the other is not. If, for instance, atomic particles obeyed this principle but gravitational fields did not, one might try to design an experiment, at least in theory, by which one would determine very precisely the values of all components of the gravitational field surrounding a massive particle, and then calculate with equal precision data about the source of the gravitational field, the particle itself.

Accordingly, a number of attempts are under way to extend quantum theory to the gravitational field. This program has not yet been fully successful, although there is no known reason for its failure. The difficulties encountered are interesting in themselves. They lie in the peculiar nature of Einstein's theory. Whereas in all earlier theories, including the special theory of relativity, space and time had been assumed to be endowed with a rigid geometric structure independent of the physical processes taking place, Einstein's theory invests space and time with little more, to begin with, than the existence of world points and of a degree of continuity sufficient to allow the assertion that some world points are "close" to each other.

Geometric structure (in the sense of a quantitatively defined distance between world points, and of a parallel displacement of directions from one world point to a neighboring point) is brought into the space-time continuum only through the gravitational field. This geometric structure is not rigid but subject to local conditions and to the laws of physics. As every kind of field carries energy, and as energy acts as a source of gravity, all fields, and all particles, contribute to the geometric structure. If the gravitational field is to be subjected to Heisenberg's uncertainty principle, then the geometry of space and time cannot be free of uncertainty, either. The precise relationships between world points must be abandoned in favor of some probabilistic formulations. For this reason, the quantum theory of the gravitational field calls for novel conceptual approaches not required in the past but which, once established, are bound to affect the whole of physics.

23 What Is an Observable?

According to quantum theory, even the most comprehensive information about the past history of a physical system does not permit certain prediction of the outcome of new observations. Instead, for each conceivable type of future measurement, the theory tells what may be the possible outcomes, and depending on past information, it associates with each of these possible outcomes a probability, or frequency, of its actual occurrence. Quantum theory has nothing to say about quantitities that cannot be observed by some experimental procedure. Physical quantities accessible to observation and measurement are referred to as *observables,* and this concept is common to quantum and to non-quantum theories.

Until the advent of general relativity, physics considered hardly any quantities that were not susceptible to observation and measurement. Moreover, at least in pre-quantum physics, it was taken for granted that, given sufficient empirical information about the past history of a system, the dynamic laws yielded precise information about the values of all observable quantities at any later time. Confirming a conjectured dynamic law was tantamount to comparing theoretical predictions of such future values with actual experimental determinations of the same quantities.

General relativity has brought about a serious complication in the concept of observable quantities, because of its adoption of the principle of general covariance. In a Newtonian physical theory, or even in special relativity, description of the past history of a physical system necessarily involves the adoption of a particular inertial frame of reference, and this frame of reference is then determined once and for all. All predictions of the future are understood to be in terms of this same inertial frame of reference; it is in this sense that a qualification such as "ten seconds from now . . ." is to be understood in the statement that continues ". . . the particle will be found at such-and-such a location."

Once the principle of general covariance has been adopted, determination of a coordinate system in the description of a system's past history by itself fails to fix the four-dimensional coordinate system to be used in formulating predictions of future behavior. This lack of unique determination was discussed to some extent in connection with the laws of motion of material bodies. At any rate, the sentence quoted in the preceding paragraph, and its qualifying clause, lacks any precise meaning in a general-covariant theory, no matter how carefully the past history of a physical system has been described. To render observable a future value of any quantity, such as a field variable, one would have to specify the precise apparatus, and its mechanical characteristics, with which the measurement is to be performed. One would, for instance, specify that such-and-such a clock, free-falling, and showing on its dial its own proper time, will collide with a certain test particle and that, at the instant of collision, the hands of the clock will indicate a certain time. In such a statement, the specification of coordinate system has been made unnecessary.

Historically, the gravitational field in general relativity has always been described in terms of the values of the individual field variables at world points which were identified by their coordinate values. From the foregoing discussion, it follows that these quantities can be neither observed nor predicted, because a world point can be identified through coordinate values only if the coordinate system right in the vicinity of that world point has been fixed in terms of other observations. Hence none of the parameters customarily used for describing a physical situation is an observable in the strict sense of that term. In any quantum theory of the general-relativistic gravitational field, these are not the quantities about whose values one can make probabilistic predictions.

But though it is meaningless to ask for the values of field variables at world points specified in terms of coordinate values, physical situations are distinct. There must be observational procedures by which one can ascertain some particulars, which will set one physical situation apart from other conceivable physical situations. Otherwise, general relativity would be empty, without relation to reality. But if there are distinguishing marks in every physical situation, if one can tell one gravitational field from another, then there must exist mathematical expressions that formalize these distinctions. Such mathematical expressions will then play the role of observables in the theory of gravitation. The quantum theory of the gravitational field must be built around them.

It is one thing to realize that there must be observables in general relativity. It is quite another to construct them in any useful manner. As a matter of fact, the technology of observables is still in a very unsatisfactory state. Several authors have advanced proposals for unearthing the observables of gravitation, but all the procedures currently known are exceedingly cumbersome. There does not appear to be any difficulty of principle, but a viable technique is still lacking.

One proposal for the construction of observables is based on the idea of *coincidences.* Though identification of a world point by means of coordinate values depends on the choice of coordinates, a point may be identified objectively by distinct properties of the local geometric structure. Suppose, for instance, that the curvature (which has as many as twenty different components) varies from point to point, then a particular world point may be identified in terms of the values of several components of the curvature then and there. At that same point there are other geometric quantities assuming particular numerical values, such as the rate of change of the curvature per unit distance in various directions. Some of these quantities are related to each other by the field laws, so that their respective values are not independent of each other. But not all these relations are fixed. In fact, at any one world point there is an infinity of geometric properties that are not determined by any universal laws: If in two distinct gravitational fields one world point in each is identified by the values of four distinct geometric characteristics, then at these two "corresponding" points in the two situations, other geometric parameters will have different values in the two manifolds. These different values serve to distinguish between the two physical situations and they are accessible to experimental determination. Hence they are observables in every sense of that term.

Another proposal is directed only at a special class of physical situations, those in which there is gravitational radiation but in which the geometry flattens out in all directions so as to assume the characteristics of the Minkowski continuum at infinity. The existence of ponderable bodies in finite numbers is not precluded. In physical situations of this type, one can describe the gravitational radiation at great distances and in all directions in terms of the wave patterns that would be recorded by suitable detectors. Because of the approximate flatness at large distances, the geometric relations between remote world points assume some of the patterns ordinarily associated with complete flatness. As a result, the mathematical task of setting up reference marks to which locations in space and time may be related is simplified, and the construction of observables resembling those of more familiar physical theories is facilitated. Whether these observables (called *news functions* by Bondi) have enough in common with those occurring in the full theory to warrant much hope for progress in the general case is as yet an open question.

24 Space-Time Today and Tomorrow

Thoughout this book, the properties of space and time have provided a background theme against which the evolution of man's understanding of gravitation has taken place. All through the history of the natural sciences, space and time have served as the scaffolding on which the dynamics of physical systems, and even of the whole universe, have been observed, described, analyzed, and interpreted. Until the twentieth century, by and large, this scaffolding was thought to exist with no relation to the dramatic events taking place on it, its properties being immutable and unaffected by the presence of matter and other physical objects. To be sure, as man's sophistication increased, the properties ascribed to space and time tended to become more complex, farther removed from "intuitive" concepts based on everyday experience.

In Newton's view, the three-dimensional continuum of space was flat; the postulates of Euclid applied to physical space: Space consisted of points arranged in a three-dimensional pattern; this pattern admitted of straight lines. Cartesian coordinates are the most natural coordinate systems adapted to such a manifold. Time was thought to be a one-dimensional continuum independent of space, with complete homogeneity along its (infinite) extension. Any instant in time could be chosen as the origin, with other instants arrayed forward (the future) and backward (the past) without end. The relationship between time and space did not admit a single state of motion to be identified (uniquely) as the state of absolute rest, but a whole class of states of motion were special in that they were free of intrinsic acceleration. Unaccelerated coordinate systems in space, coupled with uniform time scales, represented the class of inertial frames of reference.

In the special theory of relativity, the time continuum is no longer independent of the space continuum: together they form a four-dimensional continuum, the Minkowski universe, which has a well-defined (and flat) geometry of its own. The Minkowski universe, also, is not affected by the objects in it. Inertial coordinate systems remain a clearly defined, distinct class of coordinate systems, peculiarly adapted to the Minkowski geometry.

The transition from special to general relativity again involves construction of a four-dimensional continuum, which locally resembles a Minkowski universe. But the new continuum is no longer flat. In its details, the geometry now depends on the gravitational field, which in turn is affected by the distribution of its sources.

Whatever the geometry, world points are identified by their coordinate values. In Newton's space and time, and in the Minkowski universe, the relationships between several world points are completely determined if their respective coordinate values are known. This is so both because the geometries of these manifolds are fixed once and for all, and because all the "appropriate" coordinate systems, that is, the inertial frames, are essentially rigid replicas of each other, with the same right angles between the coordinate axes. Knowledge of the distance between two world points does not fix their coordinates but the converse is true: If the respective coordinate values of two world points are known, then the distance between them can be calculated, by a simple formula. Further, a Lorentz coordinate system can be fixed completely by a few simple data. For instance, it is sufficient to fix the origin of the time axis (the instant to be called "the present"), the zeros of the spatial axes at two different times (in order to determine the motion of the inertial frame), and the directions of the three spatial axes (once only) by means of three angles. These are ten individual pieces of information. Essentially the same ten pieces of information will serve to fix an inertial frame in either Newtonian physics or in the special theory of relativity.

In general relativity no straight-line coordinates exist, nor any other rigidly fixed coordinate systems. Hence no finite number of pieces of information suffice to fix any four-dimensional coordinate system. By themselves, the respective coordinates of the two world points convey no information about the distance between them, nor about any other geometric relationship between them. When gravitational fields are described in terms of the geometry imposed on the four-dimensional continuum of world points, this description might be interpreted as follows: "It is possible to find a (four-dimensional) coordinate system in which the curvature tensor takes such-and-such a form." This is indeed a valid and complete characterization of a particular geometry. But the same geometry may be identified in an infinity of alternative ways. Given one such description, an arbitrary coordinate transformation (transition to a new coordinate system, to a new frame of reference) produces a new and different description of the same geometry, which is just as valid as the previous one.

According to general relativity, the space-time continuum

still consists of world points; and these world points may be identified, though not uniquely, by coordinates. Given a particular gravitational field, world points may be identified, and uniquely, in terms of geometric properties. As there are infinitely many geometric parameters at every point, identification of points is also possible in an infinity of different ways, but these different methods of point identification can be translated into each other, not according to a universal dictionary but only in terms of a particular geometry. In fact, the "coincidences," mentioned in connection with the search for observables, represent this dictionary.

What is left of the world point? With the dissolution of the once straight and flat manifold and its conversion into a curved and buckled irregular continuum, the role of the world point has been eroded. It is not inconceivable that in the present formulation of general relativity the world point continues to exist as a relic from previous physical theories, to be discarded at the next stage of theoretical development.

Like all other theories of nature, relativity is certain to require modification, and perhaps even complete replacement, as man's actual knowledge of the physical universe increases. But though not final, every major theoretical development has contributed permanently to man's ability to conceive his surroundings. Newton's classical mechanics clarified the notions of mass and force; Faraday's and Maxwell's synthesis of the electromagnetic laws established the concept of physical fields. The theory of relativity has taught men that space and time are not an immutable backdrop to the dynamic unfolding of physical systems; on the contrary, the field of gravitation and the geometry of space-time are one. The conception of geometry as an ever-changing aspect of the real world rather than an abstract mathematical structure is a contribution that will survive the specific aspects of Einstein's laws of the gravitational field.

APPENDIXES

GLOSSARY

SUGGESTIONS FOR FURTHER READING

INDEX

APPENDIX I

The Equal-Areas Law of Kepler

Kepler discovered that in planetary motion the radius vector of a planet (the straight-line segment between the sun and the planet) sweeps out equal areas in equal times, as indicated in Fig. 2. This appendix shows that the equal-areas law will be obeyed if the acceleration of the planet at each instant is directed toward the sun.

Replace the actual orbit of the planet by a sequence of straight-line segments, as indicated in Fig. 68. (This replacement is not necessary but useful if the argument is to avoid the use of formal calculus.) The individual segments in Fig. 68 have been chosen so that the planet's radius vector sweeps through each triangle ($\overline{AOB}$, $\overline{BOC}$, . . .) in the same length of time (say, a month). Figure 69 illustrates the case of unaccelerated motion. In this figure, the segments $\overline{AB}$, $\overline{BC}$, . . . have equal lengths. In this latter case, it is easy to prove that all triangles $\overline{AOB}$, $\overline{BOC}$, . . . have equal areas. The proof is based on the formula for the area

$$A = \tfrac{1}{2}\,ah.$$

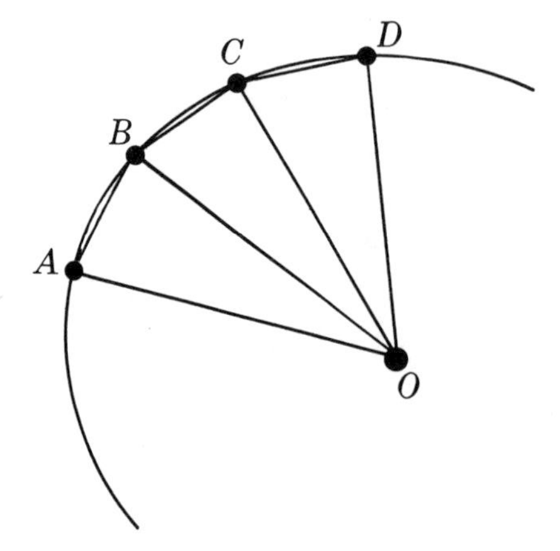

Fig. 68. Equal-Areas Law

A stands for the area of the triangle, *a* for any of its sides, and *h* for the corresponding *altitude* (the perpendicular distance of the opposite corner from the base line extending *a*). If in Fig. 69 the sides $\overline{AB}$, $\overline{BC}$, . . . are chosen as the bases, then all the triangles have their altitude in common. As the sides $\overline{AB}$, $\overline{BC}$, . . . are equal to each other, it follows that the respective areas have equal magnitudes as well.

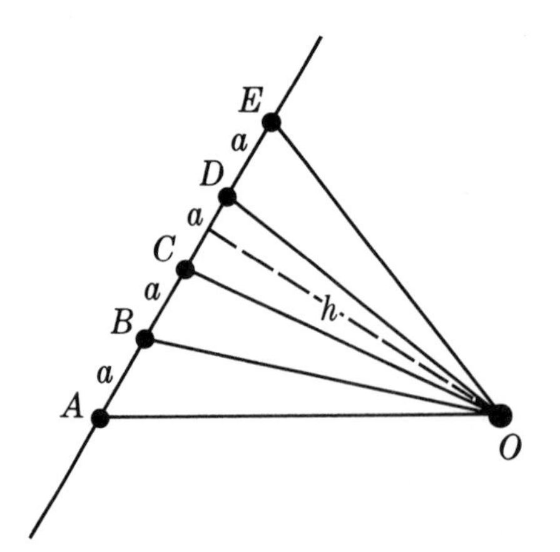

Fig. 69. Unaccelerated Motion

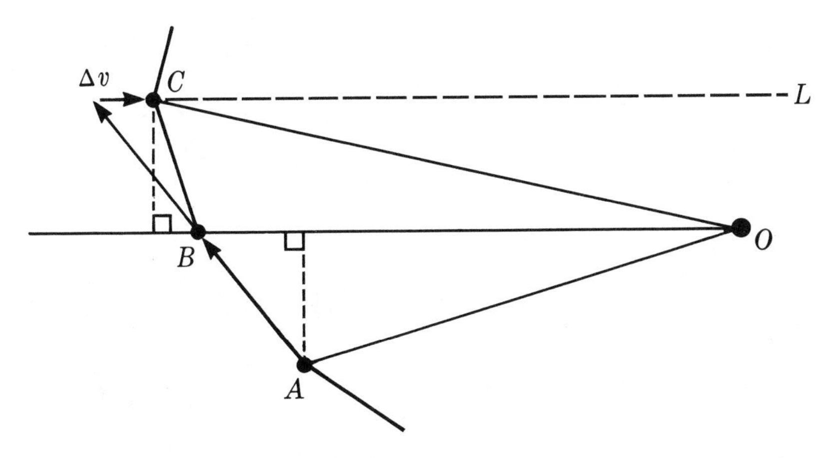

Fig. 70. Detail of Fig. 68

Return now to Fig. 68. What is the condition that the triangle $\overline{BOC}$ have an area equal to that of the triangle $\overline{AOB}$? As the side $\overline{BO}$ in one equals the side $\overline{BO}$ in the other, it follows that the perpendicular distance of C from $\overline{BO}$ must equal the perpendicular distance of A from $\overline{BO}$ (cf. Fig. 70). As the segments $\overline{AB}$, $\overline{BC}$, . . . in the further triangles represent velocities (displacements per unit time), it is to be concluded that the *change in velocity* Δv must be parallel to the radius vector BO. For sufficiently narrow triangles, (if the direction of the motion is adjusted not each month, but every day, hour, or second), the line L must pass increasingly close to the sun at O. In the limit, when points A, B, C, are no longer a finite distance apart and when the broken straight-line segments have become indistinguishable from the smooth curve of Figs. 2 and 68, the continuing changes in the direction of motion along the trajectory of the planet will point directly to the point O.

Herewith the proof is completed. The force changing the motion of the planet at every instant must be directed at the sun if the equal-areas law is to be obeyed.

APPENDIX II

Derivation of the Inverse-Square Law of Force

According to Kepler's third law of planetary motion, the *period* of a planet (the time of one complete revolution about the sun) is proportional to the 3/2 power of the length of the largest diameter of its elliptic orbit. This appendix shows that, from this fact alone, it follows that the acceleration of a planet toward the sun must be inversely proportional to the square of its distance from the sun. For simplicity, this result is derived only for circular orbits. Newton showed that the derivation holds for elliptic orbits.

On a circular trajectory, though the speed of the planet remains constant, its direction changes continuously. If the radius of the track is denoted by the symbol r, the circumference of the whole circle equals $2\pi r$. By the time the planet has traversed a small distance L, the velocity, v, has changed by the small amount w, as shown in Fig. 71. The ratio of w to v equals the ratio of L to the radius r, or

$$\frac{w}{v} = \frac{L}{r}.$$

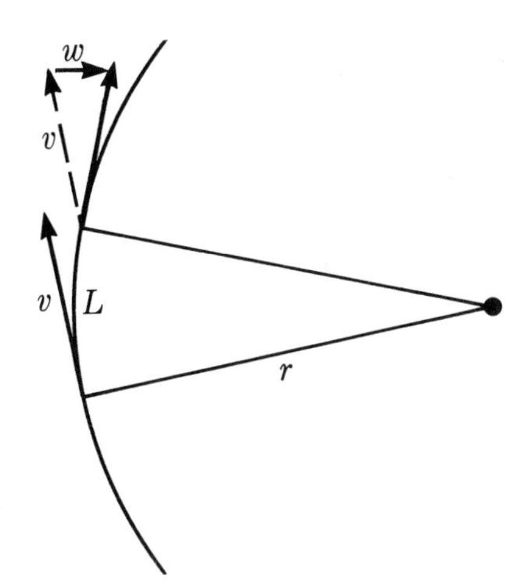

Fig. 71. Acceleration on a Circular Trajectory

If the time required to traverse the distance L be denoted by t, and the time of one period by T, then the foregoing ratios amount to

$$\frac{w}{v} = 2\pi\frac{t}{T}$$

because the speed v equals the ratio of circumference to period, or $v = 2\pi r/T$. The ratio of w to t equals the rate at which the velocity changes per unit time, or the acceleration a, so that

$$a = \frac{w}{t} = 2\pi\frac{v}{T} = 4\pi^2\frac{r}{T^2}.$$

As the force exerted on the planet by the sun equals the planet's acceleration times its mass, this force turns out to be

$$f = 4\pi^2\,\frac{mr}{T^2}.$$

If the period T is proportional to the 3/2 power of the radius r, as established observationally by Kepler on the strength of the data collected by Tycho Brahe,

$$T = br^{3/2},$$

substitution of this empirical law into the expression for the preceding force yields the desired dependence of the force on the distance of the planet from the sun:

$$f = \frac{4\pi^2}{b^2}\frac{m}{r^2}.$$

APPENDIX III

The Lorentz Transformation

Given two inertial observers in motion relative to each other, the equations that express the space and time coordinates of one in terms of the space and time coordinates of the other are called the Lorentz transformation leading from one to the other, provided that these equations satisfy the conditions imposed by the special theory of relativity. These conditions include the universality of the speed of light (in empty space). They also require that, if two events are located on a line at right angles to the relative motion of the two observers, and if they appear to take place at the same time to one observer, then they will appear simultaneous to the other observer as well, and the distance between them will appear to have the same value to both. This is true because the problematics of simultaneity uncovered by Einstein can be overcome for two events having this relationship relative to each other. If an observer occupies a location astride a line bisecting the connecting straight line, the distances between him and the respective sites of the two events will remain equal to each other at all times as long as his movements take place along this bisecting line (Fig. 72).

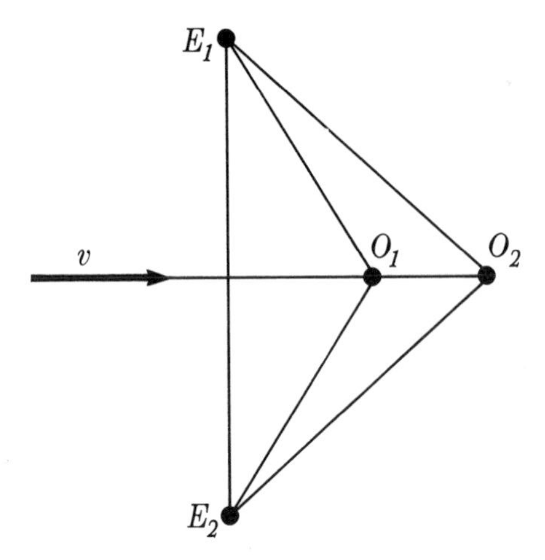

Fig. 72. Two Events on a Line Perpendicular to the Relative Motion

Assume now that the second observer is traveling relative to the first observer at a uniform rate v in the positive x direction. If his time be designated by the symbol t', his spatial coordinate in the direction of relative motion by x', and distances at right angles to the relative motion by r', a flash of light originating at the time $t' = 0$ from the location $x' = 0, r' = 0$ will propagate in all directions at the rate c. In other words, this signal will spread so that, at any time, t', subsequent to its emission, the signal will have reached in all directions the surface of a sphere of radius ct', so that the relation between time and spatial coordinates and distances will be

$$x'^2 + r'^2 = c^2 t'^2$$

or, if the terms are slightly rearranged,

$$c^2 t'^2 - x'^2 = r'^2.$$

The same signal may be observed by the other observer, whose time and spatial coordinates may be adjusted so that for him, too, the signal originates at the time $t = 0$, and at the location $x = 0$,

$r = 0$. As by assumption both observers will find the rate of propagation uniform and equal to c, the first observer will find, for the same signal being scrutinized by both of them,

$$c^2t^2 - x^2 = r^2.$$

As r represents a distance at right angles to the direction of relative motion, the two spatial distances r and r' must equal each other at all times, so that the respective space and time coordinates of the two observers are connected by the requirement:

$$c^2t'^2 - x'^2 = c^2t^2 - x^2. \tag{1}$$

This relationship must hold for all combinations x, t, x', t', as long as they refer to the same set of world points, because, for any combination of values t, x, or t', x', one can find a value of r, or r', so as to satisfy the conditions appropriate to a spreading light signal (provided that $c^2t^2 - x^2$ is not negative).

To determine the relationship between the two observers, one further assumption is required: the choice of origin of a coordinate, spatial or time, is immaterial; the relations between the coordinates are to be such that, for any two events, the coordinate differences ascertained by one observer are to determine unambiguously the coordinate differences determined by the other observer, regardless of the locations of the two events with respect to the choice of origin. Thus, the quantities t and x, or t' and x', could be interpreted less as coordinates than as coordinate differences between the world point at which the light signal originates and a world point at which it is recorded. Such direct relationships between coordinate differences are possible only if the transformation equations are linear, if the transformation equations are of the form

$$\begin{aligned} x' &= k(x - vt) \\ t' &= mt - nx \end{aligned} \tag{2}$$

with k, m, and n three constants to be determined. The first equation has been constructed so as to make certain that the origin of the primed coordinate system ($x' = 0$) has traveled the distance $x = vt$ in terms of the coordinates of the first observer while the time t has elapsed.

It remains to determine the three constants k, m, and n. If time were universal, if Newtonian physics were strictly valid, their values should be

$$k = 1, \quad m = 1, \quad n = 0.$$

These values serve as guides for the Lorentz transformation as well, for the Lorentz transformation must approach the transformation equations

$$x' = x - vt$$
$$t' = t$$

at least for relative velocities v that are small compared with the speed of light c.

Substitution of the expressions (2) into the equality (1) yields

$$(c^2m^2 - v^2k^2)t^2 + 2(vk^2 - c^2mn)tx - (k^2 - c^2n^2)x^2 = c^2t^2 - x^2$$

an equation that must hold for any combination of values of x and t. It follows that the coefficients on the left must equal those on the right; hence

$$m^2 - \left(\frac{v}{c}\right)^2 k^2 = 1 \tag{3}$$

$$vk^2 - c^2mn = 0 \tag{4}$$

$$k^2 - c^2n^2 = 1. \tag{5}$$

From these three relations k, m, and n can be determined by straightforward algebraic procedures. For instance, one can express

k in terms of n by using Eq. (5), $k^2 = 1 + c^2n^2$, and substitute into the two other equations (3) and (4):

$$m^2 - v^2n^2 = 1 + \left(\frac{v}{c}\right)^2 \tag{6}$$

$$c^2n\,(m - vn) = v\,. \tag{7}$$

The last equation may be cast into the form:

$$n = \left(\frac{v}{c}\right)^2 (m - vn)^{-1} \tag{8}$$

whereas Eq. (6) may be written in the form:

$$(m - vn)\,(m + vn) = 1 + \left(\frac{v}{c}\right)^2. \tag{9}$$

Obviously,

$$m + vn = m - vn + 2vn = (m - vn) + 2\left(\frac{v}{c}\right)^2 (m - vn)^{-1} \tag{10}$$

because of Eq. (8). If this expression for $(m + vn)$ is substituted on the left of Eq. (9), the result is

$$(m - vn)^2 = 1 - \left(\frac{v}{c}\right)^2 \tag{11}$$

which appears to leave two possibilities for the value of $(m - vn)$. Because of the need to approximate the value 1 for small values of (v/c), the negative root is, however, excluded, and the only possible value for $(m - vn)$ is

$$m - vn = \sqrt{1 - \left(\frac{v}{c}\right)^2}\,. \tag{12}$$

From Eq. (8), it follows that

$$n = \frac{v}{c^2}\left[1 - \left(\frac{v}{c}\right)^2\right]^{-1/2} \tag{13}$$

and from Eqs. (5) and (6) that

$$k = m = \left[1 - \left(\frac{v}{c}\right)^2\right]^{-1/2} \tag{14}$$

In summary, then, the Lorentz transformation equations are

$$\begin{aligned} x' &= \left[1 - \left(\frac{v}{c}\right)^2\right]^{-1/2} (x - vt) \\ t' &= \left[1 - \left(\frac{v}{c}\right)^2\right]^{-1/2} \left(t - \frac{vx}{c^2}\right) \\ r' &= r. \end{aligned} \tag{15}$$

These equations show that, with the assumptions made, a set of clocks synchronized with respect to the first frame of reference (and hence reading t-time) will not be synchronized with respect to the second frame, which is associated with t'-time. Moreover, the rates of the two sets of clocks associated with their respective frames are not equal. Each observer considers that the other's clocks are slow at the rate $[1 - (v/c)^2]^{1/2}$. He arrives at this conclusion, of course, by comparing a single clock of his colleague's with the several clocks of his own which are being passed by the "foreign" clock in the course of time.

Each observer also considers that standard scales of the other's which are aligned with the direction of motion (the x axis, or the x' axis, respectively) are contracted by a ratio equal to $[1 - (v/c)^2]^{1/2}$. Each observer measures a standard scale of the other by noting marks on his own scale that are being passed by the two respective ends of the other's scale rod at the same time (his own time).

In conclusion it might be noted that the Lorentz transformation is "reflexive": the transformation equations that result if Eq. (15) is solved with respect to x and t differ from Eq. (15) only in that the respective roles of the primed and the unprimed coordinates are interchanged, and that v is everywhere replaced by $-v$,

$$\begin{aligned} x &= \left[1 - \left(\frac{v}{c}\right)^2\right]^{-1/2}(x' + vt') \\ t &= \left[1 - \left(\frac{v}{c}\right)^2\right]^{-1/2}\left(t' + \frac{vx'}{c^2}\right) \\ r &= r'. \end{aligned} \tag{16}$$

APPENDIX IV

The Schwarzschild Radius

According to non-relativistic mechanics, the total energy of a body moving under the influence of a force field is composed of its kinetic and its potential energy; the sum of the two, the total energy, remains unchanged in the course of the body's motion.

The kinetic energy of a body depends entirely on its mass and on its speed, according to the formula,

$$K = \frac{1}{2} mv^2$$

m being the mass and v the speed. The potential energy depends on the characteristics of the force field and on the body's location within it. For Newton's inverse-square law of gravitational attraction, the potential energy of a body at a distance r from the center of a mass M is given by the expression

$$P = - \frac{\kappa m M}{r}.$$

κ, Newton's gravitational constant, is the force that two particles, each of unit mass, exert on each other across a unit distance. The total energy of a body in the gravitational field of another body is given by the expression,

$$E = K + P = \tfrac{1}{2}\, mv^2 - \frac{1}{r}\, \kappa m M .$$

The kinetic energy is always positive; the potential energy of gravitation always negative. If E is positive, the flight trajectory will lead to unlimited distances, along one branch of a hyperbola. If E is negative, the flight path is bound to the vicinity of the body with mass M, and the orbit has the shape of an ellipse. These two situations, periodic orbits versus escape to infinity, are separated by the case $E = 0$, in which the flight path has the shape of a parabola, and the body's velocity drops to zero as its distance from the center of mass, M, approaches infinity. This limiting case defines the *escape velocity*, the least speed a body must possess to manage an escape from the distance r to infinity. For $E = 0$, the velocity v is given by the expression,

$$v = \left(\frac{2\kappa M}{r}\right)^{1/2}$$

and this is the value of the escape velocity for an initial distance r from the center. For the surface of the earth, for instance, this escape velocity comes out as approximately 11.2 km $\sec^{-1}$.

The escape velocity depends on the original distance from the center of gravitational attraction, the center of the mass, M. The greater that original distance, the smaller the initial speed required for escape to infinity. Conversely, the higher the initial speed available for effecting the escape, the closer to the center the trajectory may be started. Suppose one wishes to determine the distance from which the escape velocity equals the speed of light, c, disregarding all considerations of relativistic effects. That distance then comes to be

$$R = \frac{2\kappa M}{c^2}$$

which is the expression for the Schwarzschild radius.

The calculation just made certainly has its never-never aspects. Relativity demands that no material body ever assume the speed of light. Moreover, the relativistic expression for kinetic energy at speeds approaching that of light differs drastically from the preceding non-relativistic expression. Finally, the geometry of the space surrounding a Schwarzschild mass differs markedly from flat space, so that r no longer measures distances in the ordinary sense. The only merit to be claimed for the result obtained is that even an unrealistic calculation based on incongruous theoretical assumptions may sometimes lead to the right orders of magnitude. That the non-relativistic calculation leads to precisely the expression obtained by Schwarzschild through his consistently general-relativistic treatment must, however, be considered fortuitous.

The radius of curvature of space-time in the Schwarzschild solution r_c depends on the (coordinate) distance from the center according to the expression,

$$r_c = R^{-1/2}\, r^{3/2}$$

where R is again the Schwarzschild radius. At that radius, then, R, r, and r_c are all equal. On the surface of the earth, the radius of curvature of space-time caused by the earth itself is of the order of 10^8 km, the same as the dimensions of the earth's orbit about the sun.

APPENDIX V

Gravitational Radiation

Appendix II shows that Kepler's third law leads to the inverse-square law of gravitational attraction, which was first formulated by Newton. If this law be accepted in the form,

$$a = \frac{\kappa m}{r^2}$$

where a is the acceleration resulting from the action of the body having the mass m at a distance r, and κ again denotes Newton's constant of gravitation; then it follows that, for a circular orbit of radius d, in which the centripetal acceleration is

$$a = \frac{v^2}{d}$$

the ratio between the orbital speed v and the speed of light is given by the expression,

$$\left(\frac{v}{c}\right)^2 = \frac{R}{d}$$

where R is Schwarzschild's radius, the expression obtained in Appendix III. This is also the ratio between the kinetic energy of the system and the rest energy, as the former equals the product of mass and v^2 (except for numerical factors, which are consistently omitted), the latter the product of mass and c^2.

A period of revolution will be of the same order of magnitude as the time that it takes a body with the velocity v to travel a distance of the order of the dimensions of the orbit, d,

$$T_p \sim \frac{d}{v} = \frac{d}{c}\left(\frac{d}{R}\right)^{1/2}.$$

The first factor in this expression represents the time it takes light to travel the distance, d; the second factor is the square root of the ratio between orbit dimension and Schwarzschild radius.

According to general relativity, gravitational waves have properties similar to shear waves, as explained in the section on gravitational radiation. Gravitational waves cannot be sent out by a body whose mass merely pulsates in and out without any transverse flow of matter. The source of the radiation itself must carry out some sort of shearing motion, motion that changes its quadrupole moment. The term *quadrupole moment* refers to the lack of spherical symmetry of the mass distribution as measured in terms of the deviation of the product of masses and squares of their distances from the center in a particular direction from the mean of that same expression averaged over all directions. The quadrupole moment of a double star, whose components have about equal masses, m, is of the order of md^2, d being the distance of the two components from each other.

The Newtonian field caused by a double star differs from that of a single star having the same total mass by an expression which is appreciable at distances of the order of d but drops off at large distances with the inverse-fourth power of r,

$$a_Q \sim \kappa \frac{md^2}{r^4}$$

whereas the effects of the total mass drop off only with the inverse-square power. At large distances, then, the ratio between the quadrupole correction and the principal inverse-square term goes to zero as $(d/r)^2$. These expressions are valid in Newtonian physics and neglect completely any radiative effects.

If the quadrupole moment of a star system is not constant in time but changes, for instance by rotating in space along with the double-star system, then the static effects of the quadrupole moment at large distances are overshadowed by radiative effects. The field strength of any wave, whether of gravitational, electromagnetic, or other nature, decreases at large distances with the inverse-first power of the distance. No matter how small radiative effects may be at close distances, once r is increased sufficiently, any radiation present will dominate all static effects.

Except for numerical factors, the relationship between the static expression for a_Q in the previous equation, and its value far from the source, is obtained if all but one of the factors r in the denominator are replaced by the wavelength λ, which in turn is nothing but the period of revolution T_p multiplied by the speed of propagation of the wave, c,

$$\lambda = cT_p \sim d\left(\frac{d}{R}\right)^{1/2} .$$

As a result of this substitution, a_Q at a great distance from the source will be approximately

$$a_Q \sim \kappa \frac{md^2}{\lambda^3 r} \sim \frac{c^2}{r} \left(\frac{R}{d}\right)^{5/2} .$$

The inverse-fourth-power expression for a_Q obtained for the static case holds in the dynamic case only at distances r that are small compared to the wavelength λ.

The intensity of the gravitational wave (its energy density) equals the square of the acceleration a_Q, divided by Newton's constant of gravitation

$$I \sim \frac{1}{\kappa} a_Q{}^2 \sim \frac{c^4}{\kappa r^2} \left(\frac{R}{d}\right)^5.$$

The rate of flux of energy equals the energy density multiplied by the speed of propagation,

$$p \sim \left(\frac{cR}{d}\right)^5 \frac{1}{\kappa r^2}.$$

Hence the rate at which power is radiated out in all directions is of the order

$$P \sim \frac{1}{\kappa} \left(\frac{cR}{d}\right)^5.$$

This expression permits calculating the lifetimes of the radiative process. As the mechanical energy ϵ_M of the double star is of the order of its kinetic energy, mv^2, this energy may be expressed in terms of the same quantities entering the expression for the radiated power, P, by substituting the expressions for m and for v^2 already obtained,

$$\epsilon_M \sim mv^2 \sim \frac{c^2 R}{\kappa} \cdot c^2 \frac{R}{d} = \frac{c^4 R}{\kappa} \frac{R}{d}$$

and the time in which an amount of energy of the order of the kinetic energy is lost through gravitational radiation comes out as

$$T = \frac{\epsilon_M}{P} \sim \frac{1}{\kappa} \frac{d}{c} c^5 \left(\frac{R}{d}\right)^2 \Big/ \left[\frac{1}{\kappa} c^5 \left(\frac{R}{d}\right)^5\right] = \frac{d}{c} \left(\frac{R}{d}\right)^3.$$

APPENDIX VI

Powers of 10 and Units of Measurement

For writing very large and very small numbers, both of which occur frequently in scientific discourse, it is inconvenient to have long strings of zeroes stretching either left or right of the decimal point. This inconvenience is avoided by showing the location of the decimal point by a separate factor written as a power of 10. Using the exponential method for writing all powers of 10,

$$10^2 = 100$$
$$10^3 = 1{,}000$$
$$10^6 = 1{,}000{,}000$$
$$10^{20} = 100{,}000{,}000{,}000{,}000{,}000{,}000$$
$$10^{-3} = 0.001$$
$$10^{-7} = 0.0000001$$

one can separate numbers which would otherwise be hard to read into a power of 10 and a remaining compact numerical factor. The two following instances show a large and a small number written exponentially:

$$3{,}750{,}000 = 3.75 \times 10^6$$
$$0.000{,}0047 = 4.7 \times 10^{-6}$$

The following brief tabulation presents a number of metric and other units, or standards, of length, of time, and of mass, which are used frequently in this book and in other scientific writings.

Units of Length

1 Å (Ångström unit) = 10^{-10} m
1 micrometer = 10^{-6} m
1 mm (millimeter) = 10^{-3} m
1 cm (centimeter) = 10^{-2} m = 0.4 in.
1 m (meter) = 40 in.
1 km (kilometer) = 10^{3} m = 0.6 miles
1 light second = 3×10^{8} m = 1.9×10^{5} miles
1 light year = 9.5×10^{15} m
1 parsec = 3.1×10^{16} m

Units of Time

1 hr (hour) = 3,600 sec
1 year = 3.1×10^{7} sec

Units of Mass

1 kg (kilogram) = 2.2 lb avoirdupois
1 t (metric ton) = 10^{3} kg
1 solar mass = 340,000 earth masses = 2.0×10^{33} kg

Glossary

ABERRATION: Small apparent displacement of all fixed stars from their mean locations on the *celestial sphere* occurs with a period of one year.

ACCELERATION: This is the rate at which the velocity of a material particle changes in the course of time.

As velocity itself is characterized by both magnitude (the speed of the particle) and direction, acceleration involves some change in the speed or the direction of motion, or both. When there is no change in the direction of motion, the acceleration is forward; when there is no change in speed, the acceleration is at right angles to the motion itself, as in circular motion.

Acceleration is measured in units of centimeters per second per second, or of feet per second per second or similar units. The acceleration of an automobile might be measured in miles per hour per second.

The gravitational acceleration of all objects on the surface of the earth is approximately 32 ft per sec per sec.

ACTION AT A DISTANCE: According to Newton, bodies affect each others' motions by causing accelerations, whose magnitudes also depend on the masses of the bodies being acted on. The product of acceleration and mass is called the *force* acting on a body. As this force is exerted across empty space, one speaks of action at a distance, in contrast to Maxwell's theory of electromagnetism, according to which bodies exert forces on each other only through the intermediary of the electromagnetic field.

Action at a distance is characterized by the instantaneous effect of one body on another, regardless of the distance between them, whereas action by way of fields propagates at a finite rate, which, for electromagnetic fields and for gravitational effects, is approximately 300,000 km (186,000 miles) per sec. This universal speed of propagation is denoted by the symbol *c*. (See also fields; Maxwell's theory.)

AIR SPEED: The speed, relative to the surrounding air mass, with which an airplane moves. (See also ground speed.)

CALCULUS: Newton and the German philosopher Gottfried Wilhelm Leibnitz (1646–1716) independently discovered the basic mathematical techniques by which variable quantities are treated. One such quantity is called a *function* of another (the argument), if the value of the former is determined by the value of the latter. For instance, the free fall of a body is described mathematically in terms of its location as a function of time.

Differential calculus relates the rate at which the function changes to the rate of change of the argument. The ratio of these two rates is called the *derivative* of the function. In the instance just cited, the derivative of the displacement with respect to the time is the instantaneous speed of the falling object, which is a function of time in its own right. The derivative of the speed with respect to time, in turn, is the acceleration.

Integral calculus reverses the procedure of differentiation. It is the technique of adding the increments of a function to reconstitute the original function, such as locating an object from its variable speed. Integral calculus is also used to obtain the areas and volumes of figures whose boundaries are given.

CARTESIAN COORDINATE SYSTEM: In a Cartesian coordinate system, the coordinate axes are all straight lines. They are at right angles to each other, and they are marked off in terms of the units of length adopted, such as centimeters.

Given two points with the respective Cartesian coordinates (x_1, y_1, z_1) and (x_2, y_2, z_2), the square of the distance D between them is given by the expression,

$$D^2 = (x_2 - x_1)^2 + (y_2 - y_1)^2 + (z_2 - z_1)^2.$$

CELESTIAL SPHERE: In all direct observations of extraterrestrial objects, man projects these objects on a hollow sphere, with himself as the center, whose radius has no definite meaning. This sphere is called the *celestial* sphere. Points of this sphere are usually identified in terms of two angles which correspond to the identification of points on the terrestrial globe by means of geographical latitude and longitude. On the celestial sphere these two angles are called the *declination* and the *right ascension,* respectively. According to the special theory of relativity, these angles, for the case of a very distant object (such as a fixed star), depend on the state of motion of the observer.

CENTER OF MASS; CENTROID: These are two synonymous expressions for the mean of the coordinates of an assembly of massive bodies, or particles. The averages are to be taken separately for coordinates x, y, and z in a Cartesian coordinate system, and they are to be weighted with the respective masses of the constituents.

Let the constituent particles be identified with the help of subscripts 1, 2, . . . , and their respective masses be designated by m_1, m_2, ... Then, if the respective x coordinates be x_1, x_2, . . . , the x coordinate of the center of mass is given by the expression

$$X = \frac{m_1x_1 + m_2x_2 + \cdots}{M}$$

where M stands for the total mass, $M = m_1 + m_2 + \cdots$. Corresponding expressions hold for the two remaining coordinates of the center of mass, Y and Z.

According to Newtonian mechanics, the center of mass of an assembly of particles moves on a trajectory unaffected by any forces that the constituent particles exert on each other but determined exclusively by forces originating outside the assembly. In the absence of such external forces, the center of mass travels at constant velocity on a straight-line trajectory.

CENTRIFUGAL ACCELERATION: When a body moves on a circular path at a constant speed, the direction of its velocity changes progressively; hence there is an acceleration which is usually referred to as *centrifugal* (sometimes as *centripetal*).

If the radius of the circular path is designated by the symbol r, and the constant speed by v, the time, t, for a complete circuit is

$$t = \frac{2\,\pi\,r}{v}$$

and this is the time in which the velocity vector v turns through a complete circle, whose circumference equals $2\,\pi\,v$. If the time, t, is divided into that circumference, the result is the rate at which the velocity changes, and this rate is the acceleration,

$$a = \frac{2\,\pi\,v}{t} = \frac{v^2}{r}.$$

CHARGE DENSITY; CURRENT DENSITY: These expressions refer to the amount of electric charge in a standard volume. If electric charge is measured in coulombs, charge density is expressed in coulombs per cubic meter.

As there are positive and negative electric charges, the charge density also may be positive, if positive electric charges prevail, or negative, if the opposite is true.

If the electric charges are in motion, then there is also an *electric current density*, which is the amount of charge passing through a given standard area, or cross section, per unit time. Electric current density is measured in coulombs per square meter per second. If all the electric charges present are of the same sign and move at the same speed in the same direction, the ratio between electric current density and electric charge density equals their common velocity.

CONFIGURATION: The *configuration* of a system of mass points is the total set of data referring to the location of each individual mass point in space. The velocities of the mass points do not form part of the configuration. Configuration and velocities at one instant of time, together with knowledge of their forces of interaction, represent the complete set of data which one must know in order to calculate the trajectories of all the mass points.

CONSERVATION LAWS; LAWS OF CONTINUITY: A physical quantity is said to be *conserved* when its value does not change in the course of time, if it is a constant of the motion. A quantity is also said to be conserved if it changes only under the influence of forces that originate outside the system under consideration. Among the important conserved quantities are energy, linear momentum, angular momentum, and electric charge.

A conservation law is a law of nature according to which

some quantity is conserved. If a conservation law is cast into a form in which the rate of change of the density of that quantity (the energy density, for instance) is related to the convergence or divergence of the lines of flux of that quantity, this formulation is usually referred to as a *law of continuity.*

CONTINUUM: As used in this book, the term *continuum* refers to a set of points which form a line (one-dimensional continuum), a plane (two-dimensional continuum), a three-dimensional space, four-dimensional space-time, or other manifolds. A continuum differs from a discrete set of points, in which each point is separated from all other points of the set. In a continuum, each point is surrounded by other points of the set, so that in every neighborhood of a given point, no matter how small, other points belonging to the set will be encountered.

COORDINATE SYSTEMS: To describe the locations of points or of physical objects in space, one employs systems of coordinates. These are essentially rules by which each location in three-dimensional space is identified with a triplet of numbers, and vice versa. The rules of identification must be so framed that locations near each other are characterized by small differences in the corresponding values of all three coordinates. Thus a coordinate system may be visualized as a three-dimensional grating whose points of intersection correspond to integral values of the coordinates, whereas fractional coordinate values of the coordinates represent points inside a mesh.

Only two coordinates are required to identify points on a two-dimensional surface, and only one is needed to identify points on a curve. Finally, according to Minkowski, one may think of events, which are localized both in space and in time, as points in a four-dimensional *continuum.* To localize an event in space-time four coordinates are required. (See also Cartesian coordinate system; frame of reference.)

COVARIANT; INVARIANT: A set of relationships between mathematical or physical quantities is called *covariant* if, from the validity of these relationships in one coordinate system, there follows the validity of formally identical relationships in another coordinate system. If the set of relationships is covariant under the Lorentz transformations of the special theory of relativity, the relationships, or mathematical equations, are said to be Lorentz-covariant. If the relations are reproduced under all curvilinear coordinate transformations of the general theory of relativity, one speaks of *general covariance.*

If an individual mathematical relationship, or equation, takes the same form in various coordinate systems, without regard to a larger set, it is called *invariant.* Covariance is a property of a set of several relationships, which are reproduced jointly if all of them are valid in one coordinate system. Invariance is the same property possessed by a single relationship. The distinction between the two terms is frequently blurred, and covariance and invariance are used interchangeably.

A set of quantities (rather than relations) is often called *covariant* if the numerical values of the quantities in one coordinate system may be determined by use of their *transformation law* from the values in another coordinate system. Typical examples of such covariant sets of quantities are the components of a vector and the components of a tensor. A single quantity whose numerical value is the same in all coordinate systems is termed *invariant,* or called a *scalar.*

CURRENT DENSITY. See charge density.

CURVATURE: In this book, *curvature* is used exclusively in the sense of intrinsic curvature of a manifold in which a parallel transport is defined. In its most general terms, the notion of parallel transport associates with any geometric structure at one point, an equivalent structure at another point, if the second point is reached from the first by way of a given connecting curve. A typical example of parallel transport is that of a vector, which is carried along a given point in such a manner as not to change direction or, if this is not possible, to change direction as little as possible. On the surface of a sphere, for instance, a "vector" may be considered only a line segment pointing in a direction tangential to the spherical surface, and a "parallel transport" necessarily involves adaptation to the shape of that surface.

If transport occurs along a curve that returns to the point of departure, that is, along a loop, the vector (or other geometric structure) may no longer be the same on return. If so, the underlying manifold is said to be *curved,* or to possess a *curvature.* The amount of curvature is usually expressed as the amount of deviation (for instance, the angle of deflection of the vector) divided by the area of the loop about which parallel transport was performed.

Though not used in this book, there is a second kind of curvature known in differential geometry, often called *extrinsic curvature,* which describes the extent to which a surface embedded in a three-dimensional space differs from a plane. The surface of a cone is extrinsically curved, though intrinsically flat.

DIPOLE; MULTIPOLE; QUADRUPOLE: If one of two particles near each other carries a positive electric charge and another an equal negative charge, then their combined electric charge is zero. Nevertheless, because the two particles are not coincident, together they will exert forces on other electric charges. These forces will generally not be directed straight toward, or straight away from, the pair, but at an angle. This combination of two charged particles is called an *electric dipole.* In the realm of magnetic interactions, no free magnetic poles exist, but magnetic dipoles are well known. A dipole has a magnitude, called its *moment,* which is the product of the distance between the two constituent particles and the charge of one of them, and a direction, the direction from the negative to the positive charge.

If two dipoles of equal magnitudes and opposite directions are laid side by side, the total dipole moment will be zero, but there will still be electric (or magnetic) effects in the vicinity. This combination is called a *quadrupole* (as there are four constituent single charges). Two quadrupoles, in turn, may form an octupole, and so on. All these constructions are known collectively as *multipoles.*

If a multipole changes in the course of time, the change will give rise to a spherical wave spreading outward with an angular distribution of intensity characteristic for the multipole character of the source. One then speaks of electric dipole radiation, magnetic quadrupole radiation, and the like. According to the general theory of relativity, there are no gravitational dipole waves, but quadrupole and higher multipole waves.

ELECTRIC CHARGE DENSITY; ELECTRIC CURRENT DENSITY. See charge density.

ELLIPSE; ELLIPTIC PATHS: If a cone is intersected by a plane, the shape of the figure of intersection depends on the angle of intersection. If the plane is perpendicular to the axis of the cone, the figure of intersection is a circle. If the plane is somewhat inclined, the figure of intersection is oval and is called an *ellipse.* If the plane is inclined so much as to be parallel to one side of the cone, the figure of intersection will not close but will extend to infinity. This figure of intersection is called a *parabola.* If the angle of inclination be increased still more, if, for instance, the plane is made to be parallel to the axis of the cone, the resulting figure will also extend to infinity, but its shape will differ somewhat from that of the parabola. This curve is called a *hyperbola.* The circle, the ellipse, the parabola, and the hyperbola are collectively known as the *conic sections.*

According to Newton's theory of planetary motion, a smaller mass moving under the influence of gravity exerted by a much larger mass will be one of the conic sections; which one depends on the total energy available. And indeed, the orbits described by the planets about the sun are all elliptic paths, distorted somewhat by the perturbing forces the planets exert on each other. (These are much larger than the relativistic effects, which also cause some distortion.) A planetary body with just enough energy to be able to escape to infinity, but not enough to retain a finite speed at infinity, would travel on a parabolic path, according to the theory. If the energy were still greater, sufficient for the body to escape to infinity with some of its kinetic energy left, the path would be hyperbolic.

ENERGY: In mechanics the capacity of a body to perform work is called its *energy*. Energy may be present in either of two forms, kinetic or potential. A body's kinetic energy is represented by its speed; it equals the amount of work done to build up its speed to a given value. The potential energy of a body is represented by its location in a field of force. It equals the amount of work expanded to bring it to a given location. The total energy of a body is the sum of its kinetic and its potential energy.

Energy is measured in the same units as work. A common energy unit used by engineers in Anglo-Saxon countries is the foot-pound, the amount of work required to lift a weight of 1 lb 1 ft from the ground. A metric unit is the joule, which represents the work required to move a mass of 1 kg the distance of 1 meter against a *force* which, unopposed, would cause the kilogram mass to accelerate at the rate of 1 meter per second in 1 sec. This force unit is also known as 1 newton.

EQUATION OF STATE: The physical characteristics of a gas or a liquid may be described in part by a mathematical equation which furnishes the pressure corresponding to any combination of temperature and available volume. This equation is called the *equation of state* of the substance so described. Ordinarily, the equation of state is based on data obtained in the laboratory, but it may also be obtained by theory from an atomic or molecular model believed to represent the microscopic structure of that substance. In that case, comparison of the theoretical equation of state with experimentally obtained data provides a test of the degree of realism of the model employed.

EQUIVALENCE PRINCIPLE: The principle of equivalence is one of the cornerstones of the general theory of relativity. Although there are several current formulations, this book uses the term to refer to the equality of the inertial and the gravitational masses of all objects. Both these concepts are as old as Newtonian mechanics. The *inertial mass* of an object is the ratio between an applied (external) force and the resulting acceleration; the *gravitational* mass is a measure of the strength with which a gravitating object attracts other massive bodies and with which it is, in turn, attracted by them. The equality of these two kinds of mass implies that all objects exposed to the same gravitational influences (such as to gravity on the surface of the earth) are accelerated at the same rate, as far as purely gravitational forces are concerned.

An alternative formulation of the principle of equivalence is based directly on the equal gravitational accelerations of all objects in the same gravitational field. If motions of massive bodies are referred to a frame of reference that is non-inertial, these motions are found to be accelerated even in the absence of true forces (interactions with other bodies). These *inertial accelerations* are also independent of the masses or other dynamical characteristics of the objects involved. Hence, gravitational and inertial accelerations have this independence in common, and they are, to this extent, equivalent. This formulation of the principle of equivalence must be qualified, however; inertial and gravitational accelerations may be distinguished from each other in that a purely inertial effect can be eliminated by adopting an inertial frame of reference.

ETHER: As long as electromagnetic radiation was visualized in close analogy to sound radiation, it was reasonable to conjecture the existence of a medium that would act as the carrier of the electromagnetic field and its energy, just as the atmosphere, a body of water, or the solid ground, serves as the carrier and the means of propagation of sound and of sound energy.

Although sound cannot propagate through a region of space devoid of matter, electromagnetic waves propagate with the least loss of energy, and at the highest observed rate of speed, through manifestly empty space, such as that between the celestial bodies. Hence the conjectural medium of the electromagnetic radiations, called the *ether,* or sometimes the *light ether,* was assumed to pervade all space irrespective of the presence of ordinary matter. It was endowed with peculiar mechanical characteristics in order to explain the observed properties of electromagnetic waves, particularly the absence of any longitudinal waves. The speed of propagation of light as obtainable from the laws of the electromagnetic field was to be valid in reference to the state of motion of the ether.

Attempts to observe an anisotropy of the speed of light relative to the earth could be interpreted as experiments to observe the motion of the ether relative to the earth, which was also called the *ether wind.*

With the formulation of the special theory of relativity, ether hypotheses lost much of their attractiveness, as it was realized that the consistent failure of all ether wind experiments was consistent with a new principle of relativity, a principle that foreclosed once and for all any possibility that the properties of the light ether might be discovered by experimentation.

EVENT: *Event* is the technical term for an occurrence sufficiently well localized in space and in time to qualify as a point (sometimes called *world point*) in the four-dimensional space-time continuum. An explosion occurring at an instant of time and at a well-defined site is a good example of an event.

FIELDS: In Newtonian physics, massive bodies are considered to influence each other directly across the reaches of empty space. The magnitude of this mutual attraction or repulsion for any two given bodies depends only on the distance between them. By contrast, electromagnetic interactions were found to depend on relative velocities as well as distance. Changes in distance and state of motion propagated at a finite rate (denoted by the symbol c) equal to the speed of light. Accordingly, the Newtonian concept of instantaneous action at a distance was gradually replaced by the notion of intermediate fields emanating from the site of the source and spreading gradually in all directions. In this picture, the force acting on an electrically charged particle results directly from the electromagnetic field in the immediate vicinity of the particle and only indirectly from the sources of that field, the remaining charged bodies.

In relativistic physics not only electromagnetic but also gravitational interactions are analyzed in terms of the field concept. In fact, the field concept currently dominates all thinking about interactions between particles, whatever the nature of that interaction may be.

FOCUS: When light rays enter an optical lens or system of lenses, as a parallel beam, they may converge to a point. The point of convergence is called a *focus* of that optical system. The same terminology is applied to rays reflected from a mirror or a combination of mirrors and lenses.

FORCE: *Force* is the technical term applied to any external action modifying the motion of a massive body or particle. In the absence of such force, and *relative to an inertial frame of reference,* the motion of a particle is unaccelerated; that is, it moves along a straight line and at a constant rate of speed. The presence of a force results in an acceleration, a change in the velocity, the amount of which is inversely proportional to the body's mass. The magnitude of the external force acting on a body is expressed as the product of its mass and its acceleration.

FRAME OF REFERENCE: This term applies to any system of standards to which the location of a physical object may be referred as it changes in the course of time. A frame of reference differs from a three-dimensional, spatial coordinate system in that the latter is fixed without any reference to time, whereas a frame of reference requires the fixation of spatial coordinates at all times. The location of the origin of the spatial coordinates and the directions in which the coordinate axes point must be given at every instant in time for the determination of the frame of reference to be complete. The trajectories of physical objects in the course of time are then described by giving the values of their coordinates at each instant in time.

From the four-dimensional point of view, a frame of reference is a four-dimensional coordinate system, three of whose coordinates point in spacelike directions, whereas the fourth coordinate axis is timelike. (See also inertial frame of reference; local frame of reference.)

FREE-FALLING FRAMES OF REFERENCE: See local frame of reference.

GAMMA RAYS: These high-frequency, penetrating electromagnetic rays are produced in connection with radioactive processes. Most gamma rays are of higher frequency, and hence of shorter wavelength, than X rays, electromagnetic radiation produced when electrons are violently accelerated or stopped. Gamma rays and X rays are distinguished in terms of their production rather than in terms of their frequency ranges; some gamma rays are of considerably lower frequency than most X rays employed in medical practice.

GENERAL COVARIANCE: See covariant.

GRAVITATIONAL WAVES: The general theory of relativity predicts that masses made to move rapidly back and forth will give rise to oscillatory gravitational fields which propagate at the same rate of speed as electromagnetic waves, usually referred to as the *speed of light.* Such oscillatory traveling fields are known as *gravitational waves.* No gravitational waves have as yet been observed, but instrumentation for detecting them is in the very early stages of development.

According to theory a gravitational wave passing across a swarm of particles originally at rest relative to each other would lead to shearing motions of the particles with respect to each other.

GROUND SPEED: This term applies to the speed of an airplane relative to the ground. The ground speed depends not only on the capabilities of the plane but also on the prevailing winds; a tail wind will increase the ground speed beyond the air speed of the plane, whereas a headwind will make the ground speed less than the air speed.

INERTIAL ACCELERATION: When the motion of material particles is described with respect to a frame of reference that is not an inertial frame, the particles exhibit accelerations not corresponding to forces acting on them. These are called *inertial accelerations.* Prime examples of such inertial accelerations are those observed when the chosen frame of reference rotates relative to an inertial frame. A particle first at rest with respect to a rotating frame is accelerated away from the axis of rotation; this acceleration is known as *centrifugal* acceleration. Its magnitude is proportional to the distance from the axis. If the particle is moving relative to the rotating frame of reference, then there is an additional inertial acceleration whose magnitude is proportional to the speed of the particle (relative to the rotating frame) and whose direction is perpendicular both to the axis of rotation and to the particle's instantaneous motion. This additional inertial acceleration is known as *Coriolis* acceleration.

INERTIAL FRAMES OF REFERENCE: An inertial frame of reference is a frame of reference with respect to which massive physical objects not subject to external forces are moving in a straight line at a uniform rate of speed. Given one inertial frame of reference, other frames moving uniformly and straightline and without rotation relative to the first inertial frame of reference are themselves inertial frames of reference.

INTERVAL; INVARIANT SPACE-TIME: In the geometry of ordinary three-dimensional space, any two points are at a definite distance from each other, whose value does not depend on the coordinate system in terms of which the two points are identified. If the coordinate system is Cartesian, then the square of that distance equals the sum of the squares of the three coordinate differences. In other coordinate systems, different, and usually more involved, expressions must be used to calculate the distance.

In Minkowski's space-time geometry, the invariant space-time interval plays a similar role. The square of a timelike interval between two *events* equals the square of the time coordinate difference, diminished by $(1/c^2)$ times the spatial distance squared between the two events, provided that a Lorentz coordinate system is used. Though the time coordinate difference between the two events depends on which coordinate system is used, and so does the spatial distance, the value of the invariant interval is the same in all Lorentz coordinate systems. In this respect, the space-time interval of Minkowski geometry resembles the distance of ordinary geometry. But whereas the distance between two points in space is zero only if the two points coincide (if they are the same point), the space-time interval between two distinct events will vanish if their relationship is lightlike; that is, if a light signal associated with one of the two events will reach the site of the second event just when this event occurs.

INVARIANT: See covariant.

INVARIANT INTERVAL: See interval.

LAWS OF CONTINUITY. See conservation laws.

LIGHT CONE: Given an event in space-time, the set of all directions in which a light signal can travel away from that event and of all directions in which a light signal can travel toward the event, forms a geometric figure called the *light cone* associated with that event. Those directions pointing toward the event constitute the past light cone; those pointing away from the event constitute the future light cone. If the past light cone is extended to infinity, its constituent lightlike rays can be identified with the points on a sphere which is called the *celestial sphere.*

LINEAR MOMENTUM: The *linear momentum* of a massive body is defined as the product of its mass by its velocity. In relativity, the mass in this definition is to be the relativistic mass, which itself depends on the velocity but equals the ordinary mass at rest. The importance of the linear momentum derives from its obeying to a conservation law. When two bodies exert forces on each other, the increase in the linear momentum of one as a result of acceleration is compensated by the diminution in the linear momentum of the other; hence their total linear momentum does not change. The same law holds for larger numbers of particles interacting with each other.

In Minkowski geometry, the total linear momentum and the total energy of a system of particles form the components of a four-dimensional vector, which is conserved in its entirety.

LOCAL FRAMES OF REFERENCE: Ordinarily it is understood that the coordinate systems associated with a frame of reference at various times extend indefinitely throughout space. When, occasionally, one defines the coordinate systems only in the immediate vicinity of some point, or of some trajectory, one speaks of a *local frame of reference.*

If a local frame of reference is so constructed that objects within its range of definition appear unaccelerated while engaged in gravitational free fall, it is called a free-falling frame of reference. It is in the nature of the gravitational field that a free-falling frame can not be extended unambiguously over a region in space and time in which the curvature of space-time becomes noticeable.

LORENTZ COORDINATE SYSTEM; LORENTZ FRAME OF REFERENCE: In

the special theory of relativity, an inertial frame of reference is often referred to as a *Lorentz frame of reference.* This term implies that the spatial coordinate system at each instant of time is to be Cartesian, and the units of time so chosen that the speed of light in empty space has the same value in all directions, at all locations, and at all times. In the four-dimensional language of Minkowski geometry, a Lorentz frame of reference is called a *Lorentz coordinate system.*

LORENTZ TRANSFORMATIONS: A *Lorentz transformation* is a transformation law for four-dimensional space-time coordinates that leads from one Lorentz coordinate system to another. The two Lorentz frames involved are either at rest relative to one another, or in a state of relative straight-line uniform motion without rotational motion. Their respective origins of space and time coordinates need not coincide, nor will their respective spatial coordinates be parallel to each other. A specific Lorentz transformation is identified in terms of 10 parameters: the relative velocity components (3); the angles between corresponding spatial coordinates (3); and the space and time coordinates of the origin of one coordinate system with respect to the other (4).

MANIFOLD: In mathematical terminology, the concept of space is often limited to sets of points which may be identified with sets of real numbers, their coordinates, so that to each set of n real numbers (in the case of an n-dimensional space) there corresponds exactly one point. Points lie close together if their respective coordinate values differ by small amounts. A space thus defined has local properties, which correspond to those associated with the notion of space in non-technical usage, and properties at large, which may be summarized by the requirement that the space be covered by a single coordinate grid.

When an infinite set of points everywhere has the local

properties of a space but cannot be covered by a single coordinate grid, it is called a *manifold.* The plane is a two-dimensional space, but the surface of a sphere is a manifold: If a coordinate grid is built, say, about the north pole, its extension to most of the sphere is straightforward, but on reaching the opposite pole, one must either assign to that (perfectly ordinary point) infinitely many different coordinate values, infinite coordinate values, or omit it from the coordinate description altogether. Two overlapping coordinate systems, however, will describe the surface of a sphere completely. Other manifolds include the surface of a doughnut, the Moebius strip, and the Klein bottle. As a space is a special kind of manifold (though not vice versa), the term *manifold* is often used to avoid implying that the set of points considered is a space, or is not.

MASS: See equivalence principle (see also center of mass).

MAXWELL'S THEORY OF THE ELECTROMAGNETIC FIELD: The laws of the electromagnetic field owe their definitive form to James Clerk Maxwell and are named for him. First there are the field laws: The lines of electric force originate only in positive electric charges and terminate only in negative electric charges. The lines of magnetic force form closed loops, as there are no single magnetic poles of either sign. The electric force about a closed-loop curve equals the rate of diminution of magnetic flux through the interior of that loop, whereas the magnetic force about such a loop equals the sum of the rate of increase of the electric flux and of the electric force through the interior. Then there are the ponderomotive laws: An electrically charged particle experiences a force proportional to its own charge and related to the surrounding electric and magnetic fields. The electric field produces a force proportional and parallel to it, whereas the magnetic field produces a force proportional to the particle's velocity and perpendicular both to that velocity and to the direction of the magnetic field.

NORMAL MODES OF VIBRATION: An extended solid body whose components are held together by elastic forces is capable of a large number of distinct modes of vibration in which all parts vibrate, or oscillate, in synchronous motion at definite frequencies. These are called the *normal modes* of that elastic body; each normal mode has its own characteristic frequency. The most general internal motions of which an elastic body is capable are represented by superpositions of its normal modes.

The notion of normal modes may be extended to the standing waves of an elastic string or of a volume of air, as in an organ pipe. The mode having the lowest characteristic frequency is called the *fundamental* mode; the others are known as *harmonics*. In the same sense, one speaks of normal modes of standing electromagnetic waves in a cavity whose walls are perfect conductors of electric currents.

ORBIT: See trajectory.

ORTHOGONAL: The term *orthogonal* (at right angles) is often preferred to perpendicular in the case of geometries which differ in some essential from the geometry of ordinary space, Minkowski geometry, for instance.

PARALLEL TRANSPORT: See curvature.

PROPER TIME: Proper time is the timelike invariant space-time interval between the (four-dimensional) points along the trajectory of a material object, such as a particle. Proper time is counted from an arbitrarily chosen point along the trajectory and increases in the future direction. A clock carried alongside the particle will generally indicate proper time if it is not affected by acceleration or by external influences.

QUADRUPOLE: See dipole.

RADIAN: Ordinarily angles are measured in degrees of arc; a right angle equals 90°, and the full circle 360°. The *radian* is an alternative measure of angle, which is often used in theoretical and mathematical work. It is defined as the length of arc cut out of a circle whose radius equals 1 if the apex of the angle is placed at the center. As the full circle has an arc length of 2π (if its radius is unity), the number of radians in a right angle equals $\frac{1}{2}\pi$. One radian equals approximately 57.3°.

RESISTANCE: This term applies to two phenomena: (a) the *viscous* resistance of air, or of any other gaseous or liquid medium, to the motion of solids through it, which tends to slow down that motion; (b) the resistance of materials to the flow of electric currents through them. To avoid ambiguity, this resistance is referred to as *electric* resistance. It is measured in ohms, equal to the voltage required to pass a current of 1 ampere.

REST ENERGY; REST MASS: The *rest energy* is the energy of a physical object, such as a particle, as observed in a frame of reference with respect to which it is at rest. The *rest mass* of a particle or body is its mass as determined in a frame of reference with respect to which it is at rest. According to the theory of relativity, rest energy equals rest mass, multiplied by the square of the speed of light.

SCALAR: See covariant.

SHEARING MOTION: In *shearing motion,* the different parts of a solid body, or of a fluid medium, move past each other in such a manner that there is no change in the volume occupied by any part. The density of the material does not change in the course of pure shearing motion, but there is a distortion of shape. Rotational motion, in which neither volume nor shape is altered, and compressional motion, in which the density of the material undergoes a change, must both be distinguished from shearing motion.

SHEAR MODE: A normal mode is called a *shear* mode if the motion is predominantly or purely shear motion. Other modes are called *compressional* modes. In a shear mode, the density of the oscillating body undergoes no change, but there is a distortion of shape. The earth possesses both shear and compressional modes, both of which may be activated by earthquakes.

SHEAR WAVES: Elastic disturbances propagate through extended media as waves. In shear waves, the motion of each material particle is at right angles to the direction of propagation; hence shear waves are also described as *transverse* waves. In a compressional wave, the motion is in the direction of propagation; such a wave is also known as a *longitudinal* wave. Solid media can support both shear and compressional waves; the latter propagate faster. Fluids support only compressional waves: there is no elastic resistance to shear; hence they have no shear oscillation.

SPEED: See velocity.

STERADIAN: The *steradian* is the unit of solid angle. It is defined as the area of spherical surface cut out of the sphere with unit radius by the solid angle if its apex is at the center. The full solid angle equals 4π. The solid angle subtended by a cone whose sides form an angle of b radians with the axis equals $2\pi(1 - \cos b)$.

STRESS TENSOR: *Stress* is the resistance of an extended medium or body to distortion. That part of the stress resisting compression is usually known as *pressure;* the resistance to distortion of shape is called the *shear* stress.

The total stress may be described mathematically in terms of a tensor, whose components represent the forces acting on internal surfaces of the several orientations.

TENSORS: *Tensors* are mathematical objects that represent a generalization of the vector concept. In a manifold described in terms of coordinates, tensors have components (called *tensor components*) which describe the tensor fully. As there are many different types of tensors (of which vectors form but a small class), no general statement can be made of the number of components belonging to one tensor. Nevertheless, for each type of tensor, and for a given number of dimensions of the underlying manifold, this number is fixed. Knowledge of the numerical values of all components of a tensor in one coordinate system permits their calculation in any other coordinate system, by means of rules known as the *transformation law* applicable to that type of tensor.

Some tensors are associated with the orientation of curves (tangent vectors), two-dimensional surfaces (bivectors), and higher-dimensional structures. Other tensors are associated with the metric properties of a manifold (metric tensors). In the physics of solids, the strain tensor describes the local distortion of the material, both shear and compression; the stress tensor incorporates all information about the internal forces that tend to restore the equilibrium shape. In the theory of relativity, the metric tensor, with ten components, relates the coordinate differences between two nearby events to the invariant space-time interval between them. The curvature tensor, which has twenty components, describes the deviation of the space-time continuum from flatness.

TRAJECTORY: The curve along which a particle or any other massive body travels in the course of time is called its trajectory. Thus one speaks of the trajectory of a star, of a planet or satellite in the solar system, and of an electron inside the picture tube of a television set. Complete knowledge of a trajectory implies information about both the points of space through which the curve passes and the times at which the particle passes through each point. In Minkowski space-time, the trajectory is a curve whose points are identified by the values of the four space-time coordinates.

The term *orbit* is used almost synonymously with trajectory, but sometimes it is implied that orbits are trajectories closed in space, as in the case of planets, and less emphasis is placed on the time element.

TRANSFORMATION LAWS: A *transformation law* is a rule by which the components of a vector, the components of a tensor, or similar quantities, may be determined in one coordinate system if the numerical values of the corresponding quantities are known in another coordinate system. If the components of a vector are given in one coordinate system, the vector as a directed quantity is thereby fully determined, and its components with respect to any other coordinate system may be obtained by projection. (See also covariant.)

VECTOR: A *vector* is a tensor whose components are equal in number to the dimensionality of the space or manifold in which it is defined. The prime example is the line segment that leads from one point to another and possesses both magnitude (the distance between the two points) and direction. Other examples include the *velocity* of a particle, which is defined as the displacement per unit of time, the *acceleration* (rate of change of the velocity), the *force* (which equals acceleration times mass), and the *electric field strength* (force exerted on a unit of electric charge).

VECTOR SUM: Several vectors may be "added" together by placing them end to end. The vector reaching from the tail of the first to the head of the last vector is the vector sum. Though vectors are not ordinary numbers but objects of geometry, the vector sum shares some characteristics with the sum of ordinary numbers, notably, that the sequence in which vectors are added together has no influence on the outcome of the addition. When vectors are described in terms of their components, the components of the vector sum are the sums of the respective components of the individual vectors.

VELOCITY: The rate of displacement of a physical object with respect to some suitable frame of reference is called its *velocity* relative to that frame. As the displacement is in some particular direction, the velocity is a vector, whose components are the rates of change of the three respective coordinates. The magnitude of the velocity vector is known as the speed of that object; the term *speed* no longer implies a definite direction. The speed of light, for instance, refers to the propagation of electromagnetic waves regardless of the direction in which that propagation takes place.

WEIGHT: The force experienced by a massive body under the influence of gravity is called its *weight.* Weight is not an intrinsic property of the body concerned (as mass is) but depends on its mass as well as on the ambient gravitational field. Thus, the weight of a body on the moon is only one-sixth of what it is on the surface of the earth. In everyday usage, the distinction between mass and weight is often blurred, because on the surface of the earth the field of gravity is almost constant, but in technical discourse, the two terms should not be confused.

Suggestions for Further Reading

The following list includes some non-technical works which may be helpful to a reader who wishes to study alternative presentations of the theory of relativity or to inform himself on collateral matters.

A. BOOKS ON RELATIVITY

EINSTEIN, ALBERT. *Relativity, the Special and General Theory: A Popular Exposition.* New York: Henry Holt & Company, Ltd., 1921.

The German original was published in 1917, the year after Einstein's definitive article on the general theory of relativity. I well remember my own delight when I studied this book as a teenager; reading it contributed significantly to my choice of theoretical physics as a life work.

RUSSELL, BERTRAND. *The ABC of Relativity.* Revised edition, edited by Felix Pirani. London: George Allen & Unwin, Ltd., 1958.

This work, also, is a classic. As might be expected from its author's own mathematical and philosophic specialization,

Russell's presentation, though technically competent, is more concerned with philosophical implications than Einstein's, whose principal aim is to explain relativity as a physical theory.

EINSTEIN, ALBERT, and INFELD, LEOPOLD. *The Evolution of Physics: The Growth of Ideas from Early Concepts to Relativity and Quanta.* New York: Simon & Schuster, Inc., 1938.

INFELD, LEOPOLD. *Albert Einstein. His Work and Its Influence on Our World.* Revised edition. New York: Charles Scribner's Sons, 1950.

These two books range over the whole of modern physics, the second stresses how its development can be traced directly to contributions by Einstein. Both heavily emphasize relativity.

BOHM, DAVID. *The Special Theory of Relativity.* New York: W. A. Benjamin, Inc., 1965.

As its title suggests, Bohm's book deals only with the special theory of relativity, which it treats much more completely than this work. Bohm discusses at considerable length his philosophical attitude toward scientific theorizing, and he uses the special theory of relativity as a prime exhibit to make his points.

LORENTZ, HENDRICK ANTOON; EINSTEIN, ALBERT; MINKOWSKI, HERMANN; and WEYL, HERMANN. *The Principle of Relativity.* Translated by W. Perrett and G. B. Jeffery, with notes by A. Sommerfeld. New York: Dover Publications, Inc., 1952.

This volume reprints the original papers proposing the special as well as the general theory of relativity, as published in the technical journals. Although reading them requires more knowledge of mathematics and physics than readers of this book are assumed to have, these papers convey some of the factors leading to the discovery of relativity that may not appear as clearly in secondhand accounts.

B. BOOKS ON CONTEMPORARY ASTRONOMY

HOYLE, FRED. *Frontiers of Astronomy.* New York: Harper and Row, Inc., 1955.

Like all other branches of experimental science, astronomy develops so rapidly that, in many respects, this book is already out of date, but its author approaches the problems of astronomy and astrophysics in such a fresh and unorthodox manner that reading his work is always exciting and pleasurable.

STRUVE, OTTO, and ZEBERGS, VELTA. *Astronomy of the 20th Century.* New York: The Macmillan Company, Inc., 1962.

Struve was one of the great observational astronomers. This comprehensive survey of all of contemporary astronomy is further noteworthy for its beautiful and instructive photographs and other illustrations.

C. ARTICLES ON QUASI-STELLAR OBJECTS

No book yet published includes an account of the discovery of the quasi-stellar objects, which occurred in 1963. The following articles in *Scientific American* and *Sky and Telescope* will give the reader more information about these objects whose nature is still the object of controversy. (The entries are arranged chronologically.) Basically new information may be confidently expected within the very near future.

Scientific American

GREENSTEIN, JESSE L. "Quasi-Stellar Radio Sources," *209*, No. 6 (December, 1963), 54–62.

BURBIDGE, GEOFFREY, and HOYLE, FRED. "The Problem of the Quasi-Stellar Objects," *215*, No. 6 (December, 1966), 40–52.

Sky and Telescope

GREEN, LOUIS C. "Dallas Conference on Super Radio Sources," *27,* No. 2 (February, 1964), 80–84.

———. "Relativistic Astrophysics," *29,* No. 3 (March, 1965), 145–49; No. 4 (April, 1965), 226–29.

"The Most Remote Objects Ever Identified," *30,* No. 1 (July, 1965), 16.

GREEN, LOUIS C. "Observational Aspects of Cosmology," *31,* No. 4 (April, 1966), 199–202.

MUMFORD, GEORGE S. "Brightness Variations of Quasars" (News Notes), *33,* No. 6 (June, 1967), 354–55.

Index

ABOUT THE AUTHOR

Peter G. Bergmann has been since 1950 Professor of Physics at Syracuse University, New York. He was born in Berlin, Germany, in 1915 and received his Ph.D. from the University of Prague in 1936. He was one of Albert Einstein's coworkers at the Institute for Advanced Study in Princeton, New Jersey, from 1936 to 1941, taking part in the search for a unified field theory.

At the present time, Professor Bergmann is directing major research projects in theoretical physics, primarily in relativity and gravitation, and lecturing at various universities around the world. He is the author of a classic text, *Introduction to the Theory of Relativity*, and of other textbooks on theoretical physics, as well as numerous articles for professional journals.

A Fellow of the American Physical Society and former Chairman of the Federation of American Scientists, to mention but two of the scientific societies to which he belongs, Professor Bergmann is one of the leading experts on relativity in the world today.